Jacob Lawson Wortman

The comparative anatomy of the teeth of the vertebrata

Jacob Lawson Wortman

The comparative anatomy of the teeth of the vertebrata

ISBN/EAN: 9783742805898

Manufactured in Europe, USA, Canada, Australia, Japa

Cover: Foto ©berggeist007 / pixelio.de

Manufactured and distributed by brebook publishing software (www.brebook.com)

Jacob Lawson Wortman

The comparative anatomy of the teeth of the vertebrata

THE COMPARATIVE ANATOMY OF THE TEETH OF THE VERTEBRATA.

BY

JACOB L. WORTMAN, A.M., M.D.,
ANATOMIST TO THE U. S. ARMY MEDICAL MUSEUM, WASHINGTON, D. C.

REPRINTED FROM
THE AMERICAN SYSTEM OF DENTISTRY,
1886.

THE COMPARATIVE ANATOMY OF THE TEETH OF THE VERTEBRATA.

By JACOB L. WORTMAN, A. M., M. D.

A STUDY of the dental organs of the Vertebrata is one replete with much interest when viewed from the standpoint of the naturalist. The circumstance that their modification is so intimately associated with the food-habits of the animal, being principally concerned in the prehension and comminution of the food, and that to these same habits we must look for the most powerful influences and incentives to modification in general, causes them to assume more than ordinary importance in the estimation of the philosophic anatomist who earnestly addresses himself to the problem of vertebrate evolution.

The fact, too, that the perfect condition in which they have been so often preserved in the fossiliferous strata of the earth's crust has frequently furnished the only evidence which we possess of the existence of forms long since extinct, causes them to be regarded as objects of still greater interest. When we reflect that with nothing more to guide his judgment than the dental series of an animal the expert palæontologist can, generally, not only indicate with great certainty the character of the food upon which the animal subsisted, but its general characteristics and relationships as well, even though the date of its existence be removed to a remote period in geologic history, but little surprise can be felt that so much thoughtful attention has been bestowed upon this set of organs.

No series of anatomical structures has proved of greater utility to the systematist who has endeavored to indicate the exact relationship or philogenetic history of mammalian forms than the teeth. Generally, the student who attempts to master the subject is discouraged almost at the very threshold of his undertaking by the apparently great diversity of tooth-forms to be met with in the mammalian class; but if looked at from a developmental point of view, and if a little careful attention is bestowed upon the plan of organization of the teeth of certain groups, it is not difficult to discover that there are certain central or primitive types from which it is easy to derive other related forms of dentition by simple addition, subtraction, or modification of parts already possessed.

Careful attention to this subject for several years past, with the assistance of the light which American palæontology is now able to throw upon the question, has convinced me more and more of the truth of this assertion; and I feel well assured that we are now in a position to lay down some broad principles in regard to dental evolution, at least among certain groups of the Mammalia, where they have been subjected to the greatest amount of modification.

Although there are many questions concerning the origin and details of tooth-evolution of many aberrant forms which remain to be solved, yet the discoveries which have been made in palæontology within the last twenty-five years leave scarcely a living group of animals, the development of whose teeth has progressed beyond the primitive stages, from which we have not gained some important information relative to the phases through which they have passed to reach their present condition. The possibility of reducing our knowledge of the dental structures of the Mammalia to a broad and comprehensive basis was long since recognized by Prof. Cope, to whom probably more than any one else we are indebted for a genuine philosophic insight into the forms and structure of these teeth. Scarcely less important are the contributions of John A. Ryder and Dr. Harrison Allen, whose learned researches into the probable causes of tooth-modification have marked notable stages in the progress of the subject and have opened new and interesting fields for investigation. Nor should we omit a mention of the researches of Flower, nor those of Tomes, Waldeyer, Frey, Hertwig, Magitot, and Legros, into the histology and development in later times.

Commonly, teeth are defined as hard bodies attached to the parietes of the mouth or oral extremity of the alimentary canal, whose chief function is the seizure and comminution of the food. Morphologically considered, however, they are specialized dermal appendages situated in the buccal cavity, and characterized by the presence of certain calcified tissue developed from the true derm or corium of the integument, known as *dentine*. It will be seen from this definition that the term "tooth," strictly speaking, is limited to those structures of the oral cavity which alone possess such tissue, although it is a recognized fact that to other epithelial or cuticular structures, found in many invertebrate and some few vertebrate forms, the term "tooth" has likewise been applied.

While they all subserve the same purpose, and are therefore analogous, their chief distinction consists in this—viz. in the latter, so far as they have been investigated, these organs consist of a corneous or horny substance, which is invariably derived from the more superficial epidermal layer, and is therefore *ecderonic* in origin. In the former a papilla arises from the corium, being sunk into a fold or pit, and eventually undergoes more or less calcification from its summit downward by a deposition in its substance of lime salts, forming *dentine*. The dentine thus formed is a hard, elastic substance, consisting of closely-set parallel tubuli, branching as they go, and whose crown may or may not be invested with an exceedingly hard and unyielding substance derived from the deeper layers of the epidermis, known as *enamel*. These are, then, *enderonic* in origin.

Those of ecderonic source include the so-called teeth of Annulosæ, Mollusca, Insectæ, etc. among the invertebrates, as well as the horny teeth of *Ornithorhynchus*, palatal plates of the *Sirenia*, and the horny teeth of the lampreys among vertebrates. If the term "tooth" is applicable to these structures, then we must likewise include the "baleen" of the *Cetacea* and the beaks of birds and reptiles, which by common consent are far removed from true teeth. For all such I think the term *oral armature* is preferable, from the fact that their production not infrequently depends upon the modification of organs widely different in origin.

On the other hand, those of enderonic source are found only within the limits of the Vertebrata, and range in form from the simple cone usual among fishes to the higher complex grinding organs of certain herbivorous mammals. They all agree in being developed from the corium of the lining membrane of the mouth, which is continuous with, and really a part of, the integument, invaginated at an early period. There is a possible exception in the pharyngeal teeth of fishes, which Ryder considers to be of hypoblastic origin or developed from the basement-layer of the mucous membrane of the alimentary canal, and which are practically the same as those of epiblastic origin, as far as their relation to the surface is concerned.

When we speak of teeth being modified dermal appendages, it will not be amiss to cite the evidence upon which such a generalization rests. This is best afforded by a study of the relationship and development of the dermal armature of certain elasmobranch fishes, of which the shark is a good example and furnishes us with one of the earliest, and therefore one of the most primitive, conditions of the Vertebrata.

In these fishes the defensive power of the integument is augmented by the production of numerous hard bodies in its substance, which have been termed "dermal denticles" by Gegenbaur. These structures, which are likewise known as "placoid scales," are distributed over the whole of the integument in shark-like fishes, and are ordinarily

FIG. 187.

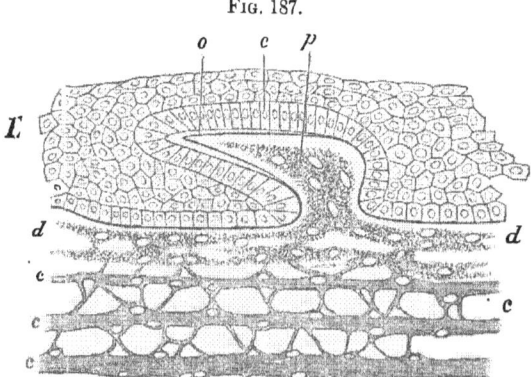

Vertical Section through the Skin of an Embryonic Shark: *c*, corium; *c, c, c*, layers of corium; *d*, uppermost layer; *p*, papilla; *E*, epidermis; *e*, its layer of columnar cells; *o*, enamel layer (from Gegenbaur, after Hertwig).

rhomboidal in form, with their apices directed obliquely backward. They consist of a solid body, which is inserted by its base into the corium, with an exposed part, which is covered with a substance indistinguishable from the enamel of the teeth. The structure of the body is likewise coincident with true dentine, and becomes fused with a basal plate of osseous material. Their development is as follows: First, a papilla arises from the uppermost layer of the corium, being covered in by the epidermis (see Fig. 187). From the deepest layer of the epidermis, or that which corresponds with the Malpighian layer, a special epithelial covering is furnished, which eventually becomes, by a process of histological differentiation, the enamel of the exposed part. The papilla, before the conversion of its substance into dentine, exhibits a central cavity, from which fine branched canals radiate to the surface. Eventually, calcification takes place, beginning at the summit, and the salts of lime are deposited in the substance of the papilla, giving rise to the dentine. Gegenbaur observes :[1] " The placoid scale has therefore the structure of dentine, is covered by enamel, and is continued at its base into a plate formed of osseous tissue; as they agree with the teeth in structure, they may be spoken of as dermal denticles."

Now, in the early embryonic stages the integument bearing these dermal denticles is pushed into the oral cavity, where they become somewhat enlarged, and appear in the adult form as teeth. Tomes says:[2] " No one can doubt, whether from the comparison of the adult forms or from the study of the development of the parts, that the teeth of the shark correspond to the teeth of other fish, and these again to those of reptiles and mammals; it may be clearly demonstrated that the teeth of the shark are nothing more than highly-developed spines of the skin, and therefore we infer that all teeth bear a similar relation to the skin." Thus the generalization is reached that teeth are but specialized dermal appendages.

With this statement of the nature of teeth in general, we are now prepared to begin a more special inquiry into the organization of a single tooth. For this purpose I have selected the third lower premolar of the dog as an average and easily-procurable example of a generalized type among the higher forms, which will serve to illustrate the composition and nomenclature of the several parts of which all teeth, with few exceptions, are made up.

For convenience of description, the several parts of most teeth can be divided into crown, fang, and neck, although there are many in which no true fangs are formed, owing to the persistent and continuous growth of the tooth; in all such no distinctions of this kind can be recognized. In the particular tooth under consideration, however, we can distinguish without difficulty an enamel-covered crown, which corresponds with the exposed part of the tooth in the recent state; two more or less cylindrical fangs or roots, by which the tooth is implanted in the aveoli and attached to the jaw bone; and a slight constriction at the point where the fangs join the crown, known as the neck (see Fig. 188). The crown in form resembles a laterally compressed cone, with an anterior and posterior cutting edge. It is covered by a dense shiny white substance of great hardness, the enamel, which ceases at the point where the fangs com-

[1] *Elements of Comparative Anatomy.* [2] *A Manual of Dental Anatomy.*

mence. At the base of the crown the enamel is thrown into a conspicuous fold or ridge, which completely encircles the tooth at this point, and is called the *cingulum*. Of the two cutting edges, the posterior is the more extensive, and is interrupted in its descent from the summit of the crown by a deep transverse notch, which constricts off a prominent cusp known as the *posterior basal tubercle*. A slight indication of a second cusp of this kind is seen immediately behind it as an elevation of cingulum. The anterior is the shorter, and descends from the apex of the crown to the cingulum without interruption. It is placed nearer the inner than the outer border of the tooth, and curves somewhat inward at its lower extremity.

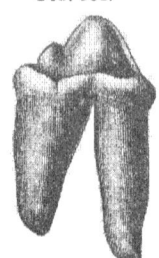

Fig. 188.

Third Lower Premolar of a Dog (*Canis familiaris*), enlarged.

The fangs are two in number, occupying an anteroposterior position, and give firm support to the crown. They are covered by a softer substance, resembling bone-tissue, known as *cementum* or *crusta petrosa* of human odontography. This material is continued over the entire surface of the crown as an excessively thin stratum in the unworn teeth of the Carnivora and several other orders, but can be demonstrated only by the most delicate manipulation and the use of the microscope. It assumes a more important relationship with the crown, as we shall presently see, in the herbivorous species of mammals.

Of the two fangs, the posterior is the larger, but the shorter, and takes the greater share in the support of the crown, although the cleft which separates them at their summits is placed directly beneath the summit of the crown. It is broad at its base, and tapers somewhat abruptly to an obtuse point. It is traversed by a vertical groove upon its anterior moiety, which fits into a corresponding ridge on the side of its socket. The anterior root is the more slender and the longer of the two. It tapers more gradually, and is likewise traversed by a broad, shallow groove upon its posterior aspect. At the point of each fang will be seen a small aperture, the *apical foramen*, through which the nerves and nutrient vessels pass to the pulp.

So far, we have spoken only of the external appearance of the tooth and of those substances which make up its outer coverings; but if both the cementum and enamel were removed, it would still preserve its original form, so great is the preponderance of the dentine as a constituent element. This can best be seen in a longitudinal vertical section, since at no part in an unworn tooth is the dentine exposed in these animals. Although the dentine is quite thick, and constitutes by far the greatest part of the tooth, it nevertheless does not form a solid body; on the contrary, a considerable cavity is hollowed out in its centre, this being largest in the part which makes up the body of the crown, and extending down each fang. This cavity lodges the dentinal pulp, the formative and nutrient organ of the tooth, and is in communication with the exterior by means of the apical foramina of the fangs.

While this structure, in common examples of enamel-covered teeth, is observable with the unassisted eye, a more minute study of the organization of the various tissues must be conducted with the aid of the micro-

356   DENTAL ANATOMY.

scope. This necessarily requires a considerable amount of experience and skill in the manipulation and preparation of material, so that to the unpractised observer a proper determination of the things which one may see is not always an easy matter. On this account I have chosen to follow the conclusions of the recognized authorities, especially the excellent treatise on dental anatomy by Charles S. Tomes, in this brief statement of the histology, rather than trust the accuracy of my own observations on the same. Since the histology of human teeth has been more fully made out than perhaps the histology of those of any other animal, it is here taken for illustration, although I am fully aware that important deviations from the structure here described are to be met with among the Vertebrata.

Dentine.—As we have already seen, the tooth consists of a dentine body with a central cavity lodging the pulp, an enamel-capped crown, and cementum-covered roots. The dentine is a hard, highly elastic, translucent substance of a yellowish-white tinge, having a silky lustre upon fracture. It is composed of an organic matrix highly impregnated with calcareous salts; through this matrix closely-set parallel tubuli radiate from the pulp-cavity toward the periphery in a direction at right angles to the surface of the tooth.

Of perfectly dry dentine the following chemical analysis is given by Von Bibra:

Organic matter (tooth-cartilage)	27.61
Fat	0.40
Calcium phosphate and fluoride	66.72
Calcium carbonate	3.36
Magnesium phosphate	1.18
Other salts	.83

The organic basis of the matrix, although closely related to that of bone, is said not to be identical with it, and is hence called "dentine" or "tooth-cartilage;" it is perfectly structureless and transparent. After the tooth has been decalcified by submitting it to the action of dilute acid for a few days, the matrix will still preserve the characteristic shape of the tooth, and can readily be studied.

As already stated, the tubuli, which are likewise known as *dental tubes*, permeate the matrix in all directions, opening freely upon the walls of the pulp-cavity, by which arrangement all parts of the dentine are brought into direct communication with the central nutrient organ, the pulp. They are most nearly approximated and their diameters greatest at their commencement on the walls of the pulp-cavity, but, pursuing a somewhat wavy course, gradually diminish in size, owing to the numerous branches which they give off. These branches, although not uniform in size, anastomose freely with those of the neighboring tubuli, and frequently show varicosities in their course. They terminate either by gradually fading out, by anastomosing with other branches, by ending in loops, or by entering the enamel and cementum layers.

While the dental tubes may be said to be channelled out in the substance of the dentine cartilage, the walls of the tubuli are not formed by this cartilage, but each tubuli is furnished with a structure known as the *dentinal sheath*, which accompanies it throughout all its plexiform radiations. The structure of these dentinal sheaths is not certainly known, owing to the impossibility of isolating them without decalcification of the dentine. Some histologists believe that they are calcified, while others express doubt as to the correctness of this conclusion. One very marked peculiarity which they possess is their great indestructibility. Dentine when submitted to the action of strong acid for a sufficient length of time to completely destroy the intervening cartilage, or when boiled in caustic alkali, will still exhibit these dentinal sheaths, for it is indeed only in this way that their presence can be demonstrated satisfactorily. One writer (Magitot) denies their existence altogether.

Enclosed within each dentinal sheath is a soft fibril, the *dentinal fibrils*, which take their origin from the cells of the odontoblastic layer of the pulp, presently to be noticed, and of which there are sufficient reasons for believing them to be nothing more than processes or prolongations. There is, however, considerable discussion upon the exact nature and relationship of these fibrils. Magitot maintains that they are continuous with a layer of reticulate cells which lie beneath the odontoblasts; these freely communicate with processes of the odontoblasts, so that there is a very direct communication between the dentinal fibrils and the nerves of the pulp. He would therefore ascribe to them a sensory function. Klein, on the other hand, holds that the odontoblasts are concerned only in the formation of the dentine matrix, and that the dentinal fibrils are long processes of deeper cells extended between the odontoblasts. Whichever of the various views now held may ultimately prevail, this much appears to be settled—viz. that the dentine is extensively invaded, so to speak, by soft plasmic material derived from the pulp, by which it is not only nourished, but also rendered highly sensitive.

In the outermost layer of the dentine, which underlies the cementum, numerous globular spaces are found, in which many of the dentinal tubes end; these are filled with soft living plasma. These spaces, if such indeed they may be properly termed, give to this layer a distinctly granular appearance, whence it was called by Tomes the "granular layer." Other structures, known as the interglobular spaces, possessing a ragged outline and short pointed processes, may frequently be seen in dried sections of dentine. They are said by Tomes to be most abundant at a little distance below the surface, and he believes them to pertain rather to a pathological than to a normal condition.

The Tooth-pulp.—It appears best to describe in connection with the dentine the *pulp* or *formative organ*, in consequence of the intimate relation which exists between them. As has already been stated, it is lodged in the pulp-cavity, and is the principal, if not the only, source of blood- and nerve-supply to the dentine. In the young and growing tooth, especially about the time calcification begins, it is largest and assumes its greatest functional activity and importance, from the fact that it is through its mediation that the dentine is formed; in fact, in the early stages of dental development, as we shall hereafter see, it is coincident with the dentine organ itself, of which in the adult tooth it is the inconsiderable remnant. As senile changes supervene it gradually loses its formative energy, and may become entirely obliterated. Taken at the adult stage of the tooth, it is seen to consist of indistinct finely fibrous connective tissue containing numerous cells. The outermost layer of the pulp is known as the *membrana eboris*, and is made up of a single layer of highly specialized cells of a dark granular appearance, somewhat elongated, termed *odontoblasts*. These odontoblasts possess large oval nuclei, and are provided with three sets of processes, as follows: the *dentinal processes*, which are identical with the dentinal fibrils, and, as we have already seen, enter the dental tubes; the *lateral processes*, by which they are connected with each other; and, lastly, the *pulp processes*, extending down to a deeper layer of cells. This latter layer of cells is somewhat intermediate in size between those more deeply seated and the odontoblasts. Three or more arteries enter at the apical foramen, and form a rich capillary plexus a short distance beneath the membrana eboris. The nerves enter by several trunks along with the arteries, and soon break up into a fine network in the substance of the pulp. According to Boll, nerve-fibres penetrate the dentinal tubuli in company with the dentinal fibrils, but this view is not fully accepted.

Cementum.—The cementum in human and many other teeth of similar structure may be said to be confined to the roots, investing them externally, unless the enamel cuticle or membrane of Nasmyth, mentioned above, pertains to it, which C. S. Tomes and others believe to be the case. It, like ordinary bone, consists of a gelatinous base combined with calcareous salts, and is permeated by vascular canals. Its histological structure presents so many characters common to bone that it is difficult to consider it anything more than a slight modification of that tissue. Just as in bone, large irregular spaces (*lacunæ*), filled with protoplasmic substance and presenting numerous minute radiating canals (*canaliculi*), which anastomose with those of neighboring lacunæ, are found in ordinarily thick cementum; certain differences are, however, seen to exist.

The lacunæ of cementum, for example, are more variable in size and are noted for the great length of their canaliculi. The direction, too, of the canaliculi is generally parallel with that of the dentinal tubuli, radiating from two sides only, whereas in bone-tissue they radiate in all directions. It has been already stated that the dentinal tubuli sometimes enter the cementum layer. When this is the case they become continuous with the canaliculi of the most deeply distributed lacunæ. The outermost or granular layer of the dentine goes so far toward establishing a complete transition in structure between the cementum and the dentine that it is generally impossible to draw a dividing-line and say where the one ends and the other begins. As to limit of distribution of the cementum on the surface of the teeth in man, monkeys, carnivores, and insectivores, different views have been expressed, owing to the various constructions that have been placed upon the nature and relationship of the enamel cuticle or Nasmyth's membrane, already mentioned. Waldeyer, Huxley, and Kölliker hold that it is no way connected with the cementum, but that it is a product derived from the enamel, and is therefore epithelial in origin. C. S. Tomes, Magitot, and Wedl, on the other hand, maintain that it is a part of the cementum extended over the entire crown of the tooth, and becomes continuous with its outermost layer in the vicinity of the neck. It is one of those excessively thin membranes (not over $\frac{1}{20000}$ inch in thickness, according to Kölliker) which are peculiarly indestructible and resist the action of the strongest acids and alkalies. When stained with the nitrate of silver, it shows a peculiarly reticulated structure resembling epithelium, which is believed by Tomes to be due to the pitted surface on its interior, by which it is applied to the enamel-prisms. Encapsuled lacunæ are likewise found in its substance, which would be difficult to explain if it were not a part of the cementum layer. Tomes has likewise traced its connection with the outer layer of the cementum on several occasions, and is therefore firmly of the opinion that it is a continuation of this tissue.

Enamel.—The excessively hard, shiny substance investing the crown of the tooth is the enamel. It is by far the hardest tissue to be met with in the animal body, being at the same time the poorest in organic constituents. Where it exists at all, it generally forms a cap of varying thickness over the exposed part of the tooth, except in those instances where there is an excessive development of cementum in this situation, which causes it to occupy a position between the cementum and dentine, as seen in the most exclusively herbivorous feeders, of which the horse, cow, and elephant are good examples. Even here palæontological evidence is quite conclusive in support of the proposition that their earlier representatives possessed teeth with naked enamel-covered crowns. This condition of nudity of the enamel is coincident with shorter cusps and less elevated ridges of the crown, and, as we have good reasons to infer from analogy, with more omnivorous habits of feeding. It can thus be shown that this anomalous arrangement of the tissues is one acquired comparatively late in the development of these forms for the exclusive purpose of giving greater strength to the lengthened cusps, thereby affording immunity from fracture during the act of mastication.

Von Bibra gives the following chemical analysis of the enamel of an adult human tooth:

Calcium phosphate and fluoride	89.82
Calcium carbonate	4.37
Magnesium phosphate	1.34
Other salts	.88
Cartilage	3.39
Fat	.20

The proportion of the organic to the inorganic material is therefore 3.59 to 96.41, while in dentine it is 28.01 to 71.99. Its structure consists of minute hexagonal prisms, known as *enamel-fibres* or *enamel-prisms*, whose long axes, broadly speaking, have a direction at right angles to the surface of the tooth. It is a comparatively rare occurrence to find the fibres pursuing a perfectly straight course from the dentine to the surface, but such is found to be the case in the enamel of the manatee or sea-cow and several other forms. Usually, they are tortuous, and frequently decussate, as in the human subject, which renders it difficult to trace the course of an individual fibre. A variety of patterns is presented by the arrangement of these prisms in the enamel of different animals, especially of the "gnawing quadrupeds," or rodents.

The prisms, when decalcified and isolated, exhibit slight varicosities or enlargements, giving them a distinct transversely striated appearance, not unlike that of voluntary muscular fibres. They are otherwise structureless. It is maintained by Bödecker that the prisms are not absolutely in contact, but that minute spaces exist between them which are filled with active protoplasmic material, which becomes continuous with that of the dentinal tubuli, thereby furnishing a means of nutrition. Some investigators admit this interstitial substance, but attribute to it no greater function than that of simple cementing material, while others, again, claim that the prisms are in absolute contact, and that no intervening substance is demonstrable. Owing to the disparity in extent between the outer and inner surface of the enamel, as well as the fact that the individual prisms do not decrease in size nor branch in their course outward to the surface, considerable spaces would be left if it were not that they are occupied by numerous prisms which do not penetrate to the dentine. The prisms end in sharp-pointed extremities which are received into corresponding pits in the enamel cuticle or membrane of Nasmyth.

DEVELOPMENT.—Next in order will be briefly noticed the development, so as to complete in this connection an entire statement of the anatomy of a single tooth. It may be said that although teeth of different types differ to a wonderful degree in their forms, which would seem to indicate differences quite as great in other respects, yet, in fact, the plan of their development is substantially the same wherever found. So far is this true that the description of the embryology of one tooth will, with little modification, answer fairly well for all teeth. The more important of these modifications in the details of development will be discussed in connection with the teeth of the various subdivisions of the Vertebrata.

We have already stated that the teeth are derived from the lining membrane of the oral cavity, which blends with the integument at the lips. The principal differences between the integument which covers the surface of the body and the mucous membrane which lines the alimentary canal are those of function and origin, the structure being essentially the same. In the one the individual cells of the epidermal layer become devitalized and scale off, while in the other they are actively engaged in the secretion of mucous, gastric, intestinal, and other juices during alimentation. The devitalization and consequent "shedding of the skin" is greater in some forms than in others. In the frogs and salamanders, for example, the skin is kept constantly moist by an abundant mucoid secretion, and the epithelium of the integument may be said to be more "alive" in these animals than in birds, reptiles, or mammals. The difference in origin consists in the important fact that the integument is formed from the epiblastic or outermost layer of primitive embryonic growth, while the mucous membrane of the alimentary canal is derived from the hypoblastic or innermost layer of the same. In the early stages of the development of the embryo the skin is more or less invaginated into the mouth-cavity, and partakes somewhat of the nature of mucous membrane proper. The real point of blending is, in the embryo at least, not at the lips, but lies inside the borders of the jaws. If, therefore, we limit the term "mucous membrane" in this situation to that tissue which is of hypoblastic origin, then the teeth of the jaws cannot be said to be developed from the mucous membrane of the mouth, as is commonly stated, but from the invaginated integument.

In many fishes teeth are found far back in the pharynx, and are attached to the gill-arches and pharyngeal bones. I am informed by Mr. J. A. Ryder, whose extensive knowledge of the embryology of fishes renders his statements highly authoritative, that these teeth lie beyond the limits of the invaginated integument, and are truly of hypoblastic derivation. If this be true, the generalization that *all* teeth are modified dermal spines is certainly incorrect. It affords us, however, an example in which identical structures have been produced from tissue of vastly different origin in a similar manner, and in all probability attributable to the same causes—viz. repeated stimulation of a particular point, which eventually gave rise to a calcified papilla.

The point at which a tooth is about to be developed is marked by a proliferation of the cellular elements of the tissue in which it will ultimately appear. These eventually arrange themselves into three organs, which have been denominated the *dentine organ*, the *enamel organ*, and the *dental sacculus*. This latter organ becomes so modified in some animals, in which coronal cement is extensively developed, as to merit the distinction of *cementum organ*. Taken collectively, they represent the tooth-germ. C. S. Tomes very justly remarks that "the tooth is not secreted or excreted by the tooth-germ, but an actual metamorphosis of the latter takes place." The three principal tissues, dentine, enamel, and cementum, thus produced, are formed from their respective organs, and consequently separate parts of the tooth-germ. Although many adult teeth do not possess enamel upon their crowns (*e. g.* edentates or sloths, armadillos, etc.), yet the presence of an enamel organ in the early stages of growth is believed to be a universal feature of the development of all teeth, and is one of the strongest arguments for their community of origin, however much they may have been subsequently modified.

The Enamel and Dentine Organs.—In the earliest stages of the development of a mammalian tooth, which is here taken for description, a slight longitudinal depression in the epithelium covering the borders of the jaws is noticeable; this is somewhat augmented in depth by the addition of a ridge upon either side of it. At the bottom of this groove the deepest or Malpighian layer of the epithelium grows down into the corium as a continuous fold or lamina, being directed downward and a little inward. In cross-section this fold resembles a tubular gland and extends throughout the entire length of the jaw. In the positions where teeth are to be formed the lower extremity of this lamina is considerably enlarged by the rapid multiplication of its constituent cells. The continuity of the fold is now broken up, and the structure which is destined to become the enamel organ appears as a process of epithelium comparable in shape to a Florence flask (Fig. 189). The outermost layer of the organ at this stage is made up of cells of the columnar variety which still retain their connection with the Malpighian layer above, from which they were orignally derived, while the interior of the enlarged extremity is composed of polygonal cells.

As development proceeds, the edges of the enlarged extremity grow more rapidly downward than the centre, which causes it to assume a bell-shaped form, with the concavity directed downward. Synchronous with this growth, a papilla arises from the corium beneath and is closely invested by the enamel organ. The appearance of this papilla marks the earliest stage in the development of the dentine organ, but it will be well to examine more closely at this stage the structure of the enamel organ. While it retained the shape of the Florence flask its periphery consisted of columnar epithelium, the interior being made up of polygonal cells. Coincidentally with its assumption of the bell shape those cells of the peripheral layer which are brought into juxtaposition with the dentine bulb or organ undergo great elongation and enlargement, forming very regular six-sided prismatic bodies, and are known as the *enamel-cells*.

FIG. 189.

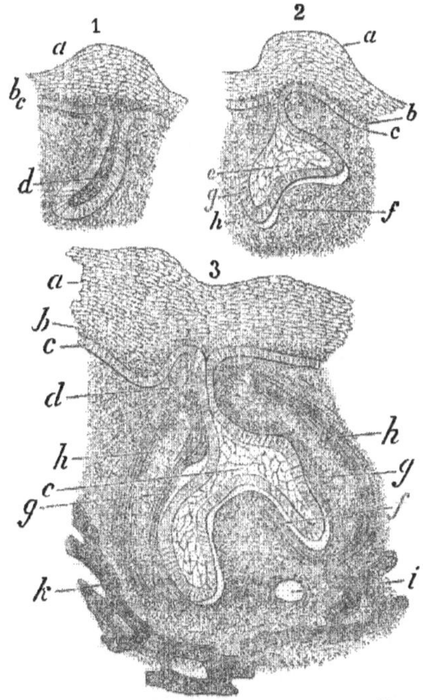

Three Stages in the Development of a Mammalian Tooth-germ: *a*, oral epithelium heaped up over germ; *b*, younger epithelial cells; *c*, deep layer of cells or rete Malpighii; *d*, inflection of epithelium for enamel germ; *e*, stellate reticulum; *f*, dentine germ; *g*, inner portion of future tooth-sac; *h*, outer portion of future tooth-sac; *i*, vessels cut across; *k*, bone of jaw (from Tomes, after Frey).

The polygonal cells of the interior are transformed into a stellate reticulum composed of cells with remarkably elongated processes; these pass through a series of unaltered cells known as the *stratum intermedium* into the enamel-cells. Lastly, we have the outer layer, which is little changed, and still remains connected with the Malpighian layer by a slender cord of epithelium. This layer is called the external epithelium of the enamel organ.

Before the dentine papilla makes its appearance "a dark halo," more vascular than the surrounding parts and corresponding to the epithelial lamina or fold which gives rise to the enamel organ, is to be seen in the submucous tissue or corium. Immediately beneath the enlarged extremity of the enamel organ the dentine papilla is developed at about the time this stage is reached by the enamel organ. In its peripheral layer highly specialized cells with several sets of processes, *odontoblasts*—already described in connection with the tooth-pulp—make their appearance, while in the remainder of the bulb numerous other cells, identical with those of the tooth-pulp, are developed. It also becomes highly vascular. Very soon the odontoblasts nearest the surface undergo metamorphosis into a gelatinous matrix, and their nuclei disappear; they are next calcified from the summit downward, and we soon recognize a thin dentine cap over the entire bulb, which gradually increases as development proceeds. The central portions of the odontoblasts remain uncalcified and form the dentinal fibrils, while the lateral processes occasion the numerous anastomoses of the dentinal tubuli and fibrils seen in the adult tooth. The dentine mass is gradually thickened by successive increments from within by a repetition of the process above described, so that it will thus be readily seen that the configuration of the dentine body, and consequently the entire tooth, is established as soon as calcification has fairly set in.

Returning to the enamel organ, we can now briefly follow its development to completion. We have already seen that it consists of an outer layer of columnar epithelium covering the convex portion, and is connected by a slender cord with the Malpighian layer above. It consists also in part of an internal stellate reticulum which passes by means of a layer of rounded cells (stratum intermedium) into the enlarged, greatly-elongated prismatic cells lining the concave lower surface, which invests the dentine organ like a cap. Before the enamel is completed the external epithelium, the stellate reticulum, and stratum intermedium disappear altogether, but before this atrophy takes place the neck or epithelial cord of the enamel organ gives rise to the tooth-germ of the permanent tooth as a diverticulum which is developed in the same way as the germ of the first or deciduous tooth just described.

The essential part of the enamel organ, or rather that which ultimately results in the formation of enamel, consists of enamel-cells. These, as we have said, become greatly elongated and assume the form of regular hexagonal prisms, which agree in shape with the calcified enamel-prisms of the complete tooth. Just as in the odontoblasts of the dentine, they are transformed into a gelatinous matrix, the nucleus disappears, and calcification begins from above, the only difference being that the enamel-prisms calcify completely, and are therefore not tubular, while in the corresponding structures of the dentine dentinal tubuli are left. Different views have been advanced in regard to the exact destination as well as the function of the several parts of the enamel organ spoken of above as disappearing by atrophy. As to the fate of the external epithelium, Waldeyer holds that after the disappearance of the stellate pulp it becomes applied to the outer surface of the enamel as the membrane of Nasmyth, which would certainly seem to be its most natural fate; but Kölliker, Magitot, and Legros claim, on the other hand, that it disappears altogether. Most authors believe that the enamel organ is devoid of vascularity, but Beal asserts that there is a vascular network in the stratum intermedium. If it be non-vascular, then it is more than probable that the pulp represents stored-up pabulum from which the requisite formative energy is derived. If vascular, it then probably subserves a mechanical purpose only, as some authorities believe.

The Dental Sacculus and Cement Organ.—So far, no mention has been made of the development of the dental sacculus. At an early period in the growth of the dentine papilla a process of the submucous tissue arises from its base and seems to grow upward on the outside of both dentine and enamel organs, finally coalescing on top, so as to enclose the growing tooth-germ in a shut sac, the dental sacculus. Whether there is an actual growth of processes from the base of the dentine bulb, or whether the adjacent connective tissue is transformed into it, appears not to have been very accurately determined; at all events, the connective tissue immediately in contact with the germ soon becomes distinguishable from that external to it by becoming richer in cells, vessels, and fibrillar elements. When the sacculus is fully formed, it is made up of an outer and an inner wall, both richly vascular. The outer wall becomes the dental periosteum, while in the inner wall, especially in the vicinity of the roots, osteoblasts appear and are calcified into cementum, as in the formation of ordinary bone-tissue. Its close application to the surface of the enamel, and partial or imperfect calcification in most teeth, give rise to the membrane of Nasmyth. In those animals, however, in which coronal cement is formed, such as the Herbivora, there is developed in connection with the inner wall, between it and the enamel, a fibro-cartilaginous structure containing characteristic cartilage-cells. These undergo calcification in a manner not different from that seen in the formation of cartilage bone, and produce the cementum in the teeth of these animals. It is then known as the *cementum organ*.

We have now made clear, we trust, as complete a statement of the anatomy of a single tooth as is consistent with brevity, but which will serve as a basis for the comprehension of the more special part of our subject—viz. the morphology of the teeth in the various subdivisions of the Vertebrata.

The Accessory Organs—the Teeth, their Structure, Development, Replacement, and Attachment, in Fishes.

It will be impossible to gain anything like a concise understanding of the dental organs of this extensive assemblage of vertebrate forms until we have first briefly outlined their classification. In this I have followed Prof. Gill, believing that his interpretations more nearly coincide with a natural arrangement.

It is a common practice of naturalists to consider the Vertebrata as divisible into five classes, as follows: *Pisces*, or fishes; *Batrachia*, or frogs, salamanders, etc.; *Reptilia*, or snakes, turtles, lizards, etc.; *Aves*, or birds; and *Mammalia*, or mammals; but according to Prof. Gill there are differences quite as great, if not greater, between certain members of the old class Pisces as there are, for example, between some fishes and frogs. For this reason he divides the permanently gill-bearing vertebrates, or those which aërate the blood throughout the entire life of the individual by means of specially adapted organs known as "gills," into four classes, which he defines as follows:

I. Skull undeveloped, with the notochord persistent and extending to the anterior end of the head. Brain not distinctly differentiated. Heart none.
LEPTOCARDII.
II. Skull more or less developed, with the notochord not continued forward beyond the pituitary body. Brain differentiated and distinctly developed. Heart developed and divided at least into auricle and ventricle.
A. Skull imperfectly developed, with no lower jaw. Paired fins undeveloped, with no shoulder-girdle nor pelvic elements. Gills purse-shaped.
MARSIPOBRANCHII.
B. Skull well developed, with a lower jaw. Paired fins developed (sometimes absent through atrophy), and with shoulder-girdle (lyriform or furcula-shaped, curved forward, and with its respective sides connected below), and with pelvic elements. Gills not purse-shaped . . . . LYRIFERA.
a. Skull without membrane bones ("a rudimental opercular bone" in *Chimæra*); gills not free, the branchial openings slit-like, usually several in number; exoskeleton placoid, sometimes obsolete; eggs few and large.
ELASMOBRANCHII.
b. Skull with membrane bones; gills free; branchial openings a single slit on each side, sometimes confluent; exoskeleton various, not placoid; eggs comparatively small and numerous . . . . PISCES.

The first of these classes, *Leptocardii*, includes a few small fish-like animals, such as the well-known amphioxus or lancelet occurring on our coast, in which no skull exists. They are in many ways most remarkable forms, being the most primitive of all vertebrates, but as they are devoid of teeth, this class can be dismissed without further consideration. The next, *Marsipobranchii*, embraces the lampreys, whose "horny teeth" have already been alluded to. The relationship as well as examples of each order of the remaining two classes is expressed in the subjoined table (p. 366), which is compiled from Dr. Gill's papers on the classification of fishes.

The Accessory Organs.—A consideration of these organs necessarily involves not only a study of the bones and cartilages taking share in the boundary of the oral cavity, but of all bones and cartilages in connection with which teeth are developed. It would likewise properly include a mention of the muscles which move these parts, together with the vascular and nervous supply; but owing to their great range of variation, as well as the limited space at my disposal, these latter will not be considered. This, in my judgment, is best accomplished by describing the normal arrangement in some typical fish and comparing all others with it. For this purpose a gadoid fish, or one of the cod tribe, is most suitable, since it exhibits the structure which obtains in a large majority of ichthyic forms.

If a well-cleaned skull be examined, it will be seen to consist, in the first place, of a cranium or brain-box, or that part which remains intact after the skull has been boiled or macerated a sufficient length of time to cause the soft parts to disappear and the arches and appendages to become disarticulated. This contains the brain, and becomes continuous at its lower back part with the vertebræ or axial pieces of the body skeleton into which the spinal cord passes. Suspended from either side of its posterior portion there is a chain of bones which extends down beneath the throat and bears the pectoral fins; this is known as the *shoulder-girdle* or *scapular arch* (see Fig. 190).

A short distance in front of this, or at a point about midway between the root of the scapular arch and the eye-socket, another arch springs

Class, LEPTOCARDII: example, lancelet.
Class, MARSIPOBRANCHII: ex. lampreys.

Class, ELASMOBRANCHII . . . . . . Orders, { Sub-class, HOLOCEPHALI: ex. chimera.
{ Sub-class, PLAGIOSTOMI . . { *Raiæ*: ex. rays, sawfishes, and torpedos.
{ *Squali*: ex. sharks.

Sub-class, GANOIDEI . . . . . . Super-orders,
{ Hyoganoidei . . . . . . Order, { *Cycloganoidei*: ex. bowfin.
{ *Rhomboganoidei*: ex. bony gars.
{ Brachioganoidei . . . Order, { *C. ossopterygia*: ex. polypterus.
{ Dipnoi . . . . Order, { *Sirenoidei*: ex. mudfishes.
{ *S luchostomi*: shovel-nose sturgeon.
{ Chondroganoidei . . Order, { *Chondrostei*: sturgeons.

Class, PISCES . . . . Sub-class, TELEOSTEI . . Orders,
{ *Opisthomi*: spiny eels.
{ *Apodes*: ex. eels.
{ *Symbiophori*.
{ *Neuatognathi*: ex. catfishes.
{ *Teleocephali*: ex. carp, herring, salmon, pike, perch, etc.
{ *Pediculati*: ex. batfishes, anglers, etc.
{ *Lophobranchii*: ex. sea-horses.
{ *Plectognathi*: puffers, foolfishes, etc.

TEETH OF THE VERTEBRATA. 367

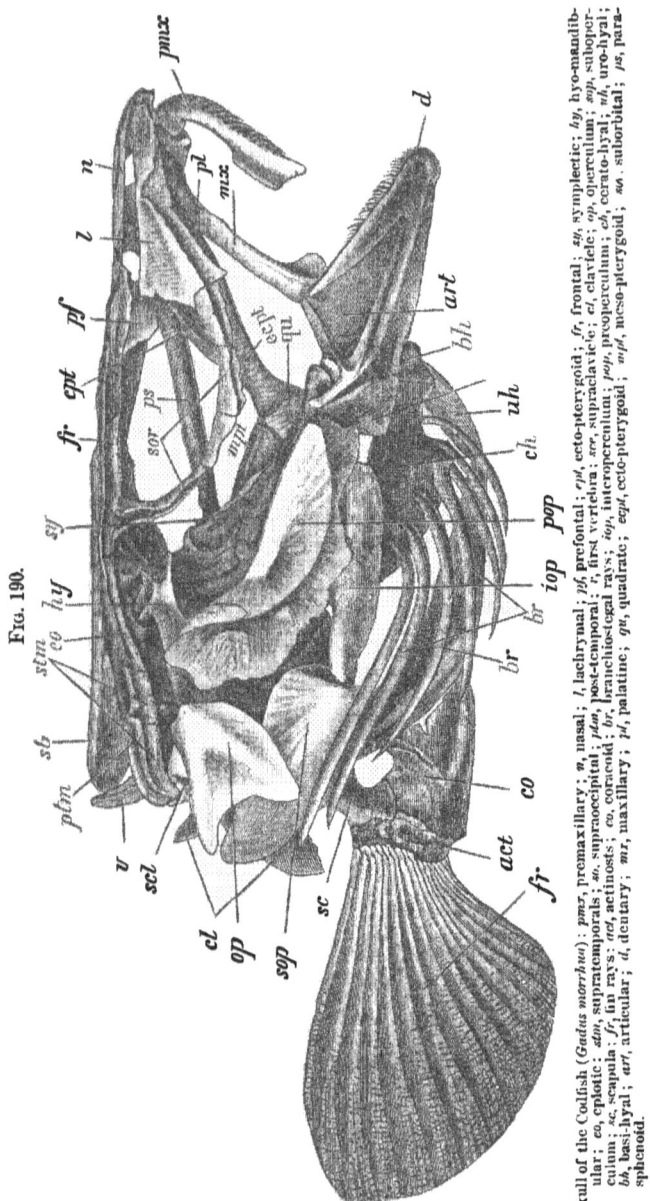

Fig. 190.

Skull of the Codfish (*Gadus morrhua*): *pmx*, premaxillary; *n*, nasal; *l*, lachrymal; *pf*, prefontal; *epl*, ecto-pterygoid; *fr*, frontal; *sy*, symplectic; *hy*, hyo-mandibular; *eo*, epiotic; *stm*, supratemporals; *so*, supraoccipital; *ptm*, post-temporal; *r*, first vertebra; *scr*, supraclavicle; *cl*, clavicle; *op*, operculum; *sop*, subopercurium; *sc*, scapula; *fr*, fin rays; *rct*, actinosts; *co*, coracoid; *br*, branchiostegal rays; *iop*, interoperculum; *pop*, preoperculum; *ch*, cerato-hyal; *uh*, uro-hyal; *bh*, basi-hyal; *art*, articular; *d*, dentary; *mx*, maxillary; *pl*, palatine; *qu*, quadrate; *ecpl*, ecto-pterygoid; *mpf*, meso-pterygoid; *so*, suborbital; *ps*, parasphenoid.

from the side-wall of the cranium and passes downward and forward to the proximal portion of the lower jaw, which is attached to it; this is known as the *hyo-mandibular arch* (see Fig. 204). Attached to the posterior portion of this arch are several broad, flat, scale-like bones which cover the gills and are called opercular bones. The upper posterior one is the *operculum*. The one in front of this, presenting a curved outline anteriorly and a posterior serrate border, is the *preoperculum*, while the two beneath are the *interoperculum* and *suboperculum* respectively. The arch itself is composed, first, of the *hyo-mandibular* bone (Fig. 190, *hy*), which by its proximal extremity is attached to the side-wall of the cranium, being lodged in a distinct oblong socket; secondly, of the *quadrate* (*qu*, Fig. 190), which articulates with it by suture at its lower extremity; thirdly, the *symplectic*, a small splint occupying a groove in the inner side of the quadrate; and, lastly, the lower jaw, which is movably articulated with the quadrate and which normally supports teeth. Each half is made up of the *dentary* or tooth-bearing piece, meeting its fellow of the opposite side in the median line or symphysis, and an articular piece which connects the dentary with the quadrate. To this may be added the *coronoid*, a small bone superimposed above the junction of the articular and dentary, and an *angular* which lies just beneath the articulation of the quadrate and articular.

From the region of the quadrate another chain of bones extends upward, forward, and inward to the anterior part of the roof of the mouth, where it is attached by ligament to the side of the vomer, or that bone which forms the prominent rostrum of the cranium after the removal of the arches. This chain is known as the *palato-quadrate* arch, and the bones entering into its composition are the *ento-*, *meso-*, *ecto-pterygoids* and the palatine. The ento-pterygoid is applied to the hyo-mandibular and quadrate upon their anterior margins; the meso-pterygoid starts out from the quadrate and ento-pterygoid, and extends toward the vomer, where it meets the palatine, which completes the arch. The ecto-pterygoid lies above the junction of the meso-pterygoid and the palatine (Fig. 190).

Immediately in front of the vomer, and attached to it and to the palatines, are two considerable bones which project downward and backward, bounding the upper posterior portion of the canthus of the mouth—the *superior maxillaries*. In front of these, again, are the *pre-* or *intermaxillaries*, limiting the anterior boundary of the oral cavity above. Another bone, which in some forms (ex. catfishes) reaches the roof of the mouth, needs to be noticed in this connection. The suborbital ring, or those bones which encircle the orbit below, articulates by its most anterior piece (lachrymal) with a bone suturally united to the cranium and taking part in the boundary of the orbit in front and above. This is the *prefrontal*, and, as already remarked in the catfishes, owing to the width of the mouth takes part in the formation of its bony roof, and in some species bears teeth. This bone is frequently mistaken for the vomer, but, as I have recently ascertained, is certainly the prefrontal, which must likewise be added to the category of tooth-supporting bones in fishes.

The several arches and bones so far enumerated, with the exception of the scapular arch—which never, to my knowledge, is dentigerous—are in direct relation with the mouth, and are exclusively concerned in prehensile and crushing functions; but those which are to follow, especially the branchial arches, were primarily used in connection with respiration, so that any relations with the teeth which they may have subsequently acquired must be looked upon as a secondary modification. This peculiarity, moreover, is of such wide application in the class *Pisces* that a description of these parts cannot well be omitted in a consideration of the accessory organs.

The hyo-branchial skeleton lies beneath the base of the cranium, and is pretty well concealed in a side view by the opercular, hyo-mandibular, and quadrate bones. It is connected with the rest of the skull at two points—viz. by the articulation of the stylo-hyal bone with the hyo-mandibular, and the other by means of loose connective tissue which binds the upper portion of the branchial arches to the base of the cranium. Its general structure will be best understood by describing it as composed of a series of transverse bony arches placed one in front of the other, rising up from the floor of the mouth and meeting in the median line above.

The most anterior of these is the *hyoid arch*, which is formed by two median basilar pieces upon either side, the *basi-hyals*. Passing from within outward, we have first the *cerato-hyals*, to which are appended the *branchiostegal rays*. The next piece in the arch is the *epi-hyal*, following which is the *stylo-hyal*. This latter bone is a slender rod, and serves to complete the connection of the hyoid arch with the hyo-mandibular bone. From the interval between the two most anterior basi-hyals there projects a small bone forward which supports the tongue, and is hence called the ento- or *hyo-glossal*. Projecting backward from the inferior surface of these same basi-hyals is another piece, the *uro-hyal*.

Behind the hyoid, and similarly composed, are the five branchial arches, of which the last two are somewhat modified. The three anterior ones are made up of median basilar bones, the *basi-branchihyals*. With these are articulated the *hypo-branchials* upon the outside, after which follow the *cerato-branchial* and *epi-branchial* pieces. In the fourth branchial arch, counting from before backward, the hypo-branchials are absent, and the uppermost segments are considerably dilated and support teeth; they are then known as the superior pharyngeal bones. The fifth arch is quite rudimentary, containing only the cerato-branchial elements, which are generally much enlarged and bear teeth; these are the inferior pharyngeal bones.

The arrangement here described is found without substantial modification except as regards relative size and the degree of ossification of the several parts in nearly all the sub-class *Teleostei*. In the *Elasmobranchii*, however, the skull remains largely cartilaginous, and the hyo-mandibular arch is always more or less imperfectly represented. The maxillæ and premaxillæ are likewise absent. In the chimeroid division (*Holocephali*) neither the hyo-mandibular nor quadrate elements can be made out, the mandible being attached directly to a broad triangular cartilaginous lamella which stretches out from the sides of the base of the skull, and whose anterior part bears the teeth of the upper jaw. It will thus be readily understood that this cartilaginous plate, continuous with the chondro-cranium, represents both the undifferentiated upper portion of the hyo-mandibular and all of the palato-quadrate arches.

In the *Plagiostomi* (sharks and rays), on the other hand, a separate cartilaginous element representing the hyo-mandibular bone is always present, and affords an articular surface to the lower jaw or mandible, which, moreover, in all the elasmobranchiates consists of a single cartilaginous bar, the primitive Meckelian cartilage. The palato-quadrate arch is likewise present and forms the dentigerous border of the upper jaw (see Fig. 204). Since the hyo-branchial skeleton in these forms is not concerned in the support of teeth, it can be dismissed without further mention.

The principal deviations in the structure and relationship of the dentigerous apparatus from the typical teleostean one to be met with in the sub-class *Ganoidei* are furnished by the *Dipnoi* and *Chondroganoidei*. The former of these orders includes the three living genera *Ceratodus*, *Protopterus*, and *Lepidosiren* of Australia, Africa, and South America respectively. They are most remarkable and interesting representatives of types in some respects low down in the scale of ichthyic organization, while in others high, in that they furnish many transitional characters between true fishes and the Batrachia (frogs and salamanders). It is highly probable that from some as yet undiscovered relative of this group the Batrachia have been derived by descent.

In this order the skull is devoid of both maxillæ and premaxillæ, and, as in the chimeroid elasmobranchiates, the hyo-mandibular arch is not completely differentiated, the lower jaw being articulated directly to the cranium. There is, however, a well-defined palato-quadrate arch supporting teeth. The hyo-branchial skeleton, although resembling the teleostean type of structure considerably, is edentulous. In the *Chondroganoidei* (sturgeons) the skull as well as the arches remain largely cartilaginous. The suspensorium (proximal part of the hyo-mandibular arch) presents two elements, usually homologized with the hyo-mandibular and quadrate pieces of the teleostean skull; the latter of these pieces affords attachment to the mandible. There is also a palato-quadrate arch. Only one species of this group, the shovel-nose sturgeon, possesses teeth, and these, according to Owen, appear only in the young.

The remainder of the *Ganoidei* agree with the *Teleostei* in the structure and arrangement of the accessory organs. The latter sub-class, however, exhibits numerous minor variations, which are confined principally to modifications of the hyo-branchial skeleton, such as the loss or atrophy of certain of its component elements; these are so numerous and varied in their nature that it would be impossible, and quite foreign to the object of the present article, to enumerate them.

Teeth of the Elasmobranchii.—As already observed, this class is divisible, not only by the differences which obtain in the arrangement of the several arches, but by the disposition, structure, and manner of replacement of the teeth, into two primary groups, of which the sharks and rays constitute one, and the Chimæræ the other. Of these, the former is the more primitive, and in all probability gave origin to the typical fishes, while the latter resembles more closely the dipnoans, and may indeed prove to have been their ancestors.

The teeth of the sharks are always numerous, and are pre-eminently adapted to the predaceous habits of their possessor. They are borne upon the cartilaginous mandibuli and palato-quadrate arches, being attached not to the cartilages themselves, but to a thick, dense fibrous membrane which forms an external investment. They are arranged in concentric rows on the summit and inner surface of the jaws, being developed from the bottom of a longitudinal fold of the lining membrane in this situation, known as the *thecal fold*. The teeth of the uppermost row, or those occupying the margins of the jaws, stand upright and do service as the functional ones until discarded; those of the next row, as well as all the succeeding ones, usually occupy a recumbent position, with their apices directed downward or upward according as they belong to the upper or lower series; but it not unfrequently happens in some species that the second, and even the third, rows may exhibit different degrees of erection. As a general rule, but a single row of teeth are in use at one time. The individual teeth composing the longitudinal rows may be disposed with reference to those of the succeeding ones so as to be parallel vertically, as is well exemplified in the genus *Lamna*, or they may be placed in such a manner as to alternate with each other, a condition seen in the blue shark (*Carcharias*). As would naturally be surmised from this arrangement, the way in which succession takes place is for the row beneath to rise up and take the place of those in use. This is accomplished by the fibrous gum in which their bases are imbedded sliding bodily over the curved surface of the jaws from within outward, continuously bringing fresh rows into position, as was long since demonstrated by Prof. Owen.

It thus happens, on account of this peculiar and, in my judgment, remarkably primitive manner of succession, that large numbers of teeth little worn are cast off during the life of each individual, and that replacement goes on far in excess of the actual requirements of the animal, and quite independently of their temporary use as organs of prehension and mastication—a fact which in itself demonstrates their dermal relationship. The only assignable cause for this extravagant development of teeth, it appears to me, is due to inequalities in the rapidity of growth in different parts of the body, which causes the integument invaginated during embryonic development to be restored or evaginated during adult growth. If this hypothesis be correct, then the whole question of the force concerned in the succession of the teeth is reduced to the simple explanation of inequalities of growth primarily, however much it may have been subsequently complicated and obscured in the higher forms. Looked at from this standpoint, it is not such an inscrutable mystery as C. S. Tomes and others would have us believe.

Considerable variety of form exists in the teeth of the different species; they may be *heterodont* (that is, different in various parts of the jaws); *isodont* (alike throughout); or *hemihomodont* (in which the individual teeth of the lower jaw are alike, but different from those of the upper jaw, and reciprocally). In all, the teeth nearest the back part of the mouth are smaller than those in front. The simplest form to be met with is the unmodified cone with a sharp point and a broad base. Such is found in the large *Rhinodon* and some "dog-fishes;" to this may be added basal denticles, as in the genus *Lamna*; or it may have a compressed triangular outline with serrate edges, as in the upper teeth of the blue shark (*Carcharias*). These lateral serratures may become so strongly developed as to give to the tooth a distinct comb-like appearance—*e. g.* lower teeth of *Notidanus* (Fig. 191).

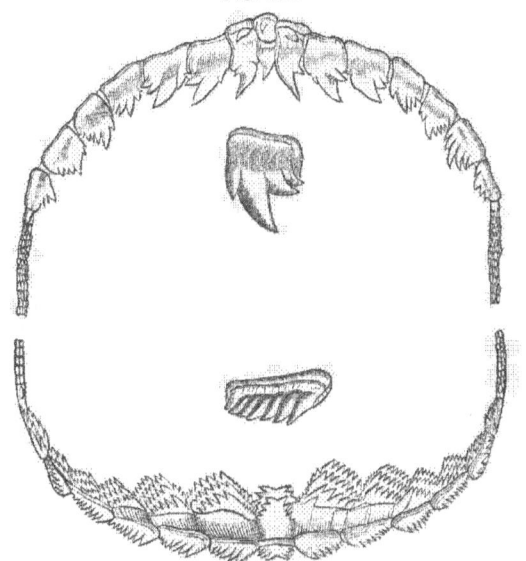

Teeth of *Notidanus* (after Gunther).

The most remarkable modification in the dental organs of sharks is exemplified by the Port Jackson shark (*Cestracion*), in which the posterior teeth gradually become broad and form a regular pavement on the surface of the jaws similar to that seen in many rays. This structure exists in consonance with the shellfish-feeding habits of the animal, in the exercise of which great crushing and comminuting power is required to be exerted. These fishes are of especial interest, inasmuch as they are the only living representatives of an extensive and widely-distributed group which appeared on the earth far back in the Devonian Epoch, and whose remains, as Owen justly remarks, " would have been scarcely intelligible to us unless the key to their nature had been afforded by the teeth and spines of the living cestracionts."

The teeth of the anterior part of the jaw (Fig. 192) are the smallest, and present a compressed conical form with the apex produced into a sharp point. Proceeding backward, they gradually assume an oblong oval outline, progressively increasing in size, their sides becoming applied to each other in such a manner as to form a regular pavement. The maximum size is attained at about the fourth tooth from the posterior end of the series, after which they decrease rapidly, although still preserving their modified crushing form.

The progressive changes in size and form, as well as the disposition, of the most highly modified teeth in this animal, are seen to be in direct accord with the uses to which they are put, and serve to illustrate, as so

TEETH OF THE VERTEBRATA.

many other dentitions do, the reasonableness of the view originally proposed by J. A. Ryder, to the effect that mechanical causes have been largely instrumental in bringing about the modifications of the teeth. It will be readily understood that the greatest mechanical advantage would be gained and the greatest pressure exerted by passing the morsel to be crushed to the posterior part of the mouth. The teeth in this situation or in its vicinity have sustained the greatest amount of strain, and are consequently most modified, while those of the anterior part of the mouth have been largely exempt from such influences, and are therefore little modified. I will have occasion to recur to this hypothesis on a future page.

The teeth of the rays present quite as great, if not a greater, range of variety than do the sharks. In general, they are more numerous, more closely crowded together, and possess forms better adapted for crushing than for seizing and lacerating. They are developed in the same way as in sharks, rising up from the bottom of a thecal fold on the inner surface of the jaw and being carried upward by a rotation outward of the

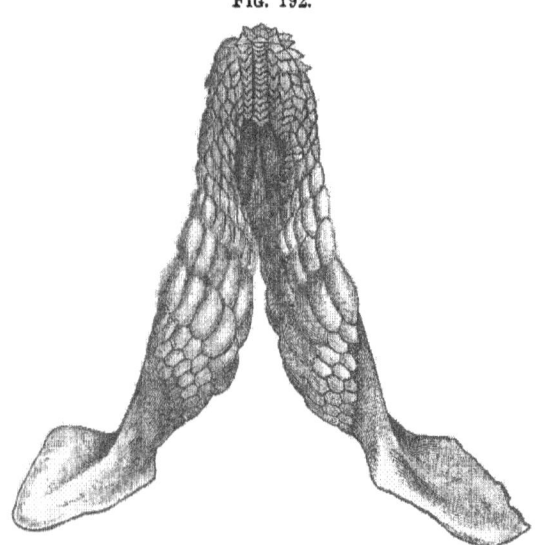

Fig. 192.

Lower Jaw of Port Jackson Shark (*Cestracion phillippsi*).

membrane in which they are imbedded. In *Raia stellulata*, from the California coast, the teeth succeed one another vertically, as in *Lamna* among the sharks, and do not form a close pavement on the biting surface of the jaws, they being separated from each other by slight intervals. In form the base of the crown represents an equilateral triangle, with the apex directed forward; from this a prominent ridge passes backward across the middle line of the base, and is produced into a sharp conical point. The teeth of the anterior part of the mouth are the largest, and gradually decrease in size as the canthus or angle of the mouth is reached. In the "barndoor skate" (*Raia lævis*) the teeth

DENTAL ANATOMY.

are more closely set, but are not in absolute contact; as in *Raja stellwata*, those of the several rows are arranged vertically, but their bases are more rounded, with only a faint indication of the backwardly projecting cusp, which is confined to the teeth of the anterior part of the jaws. In the common "stingray" (*Trygon centrurus*) the teeth are somewhat quadrangular, and have their sides directly applied to each other, forming a dental sheath of continuous pavement over the working surface of the jaws; those of the successive rows are disposed diagonally. Their crowns are of an oval form, well adapted for crushing and grinding hard substances. The "eagle rays" or "sea-devils" present a series of modifications of the teeth which diverges from that of the stingrays, and terminates in the most unique of all dentitions to be found amongst the Vertebrata—viz. that of *Aëtobatis*. Of this group the genus *Rhinoptera* possesses tessellated teeth with flat hexagonal crowns, of which

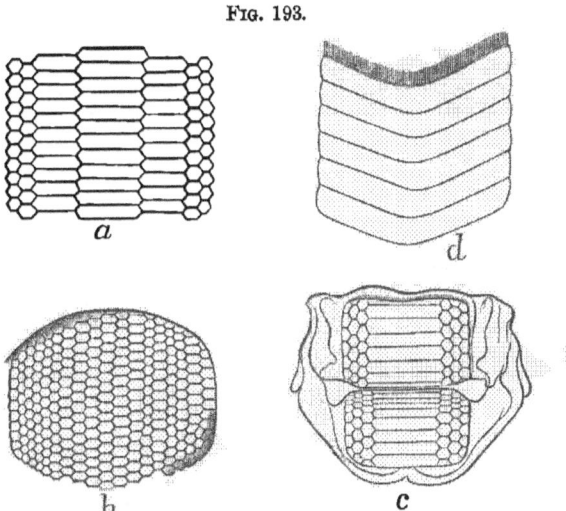

Fig. 193.

Teeth of Rays: *a*, *b*, *Rhinoptera*; *c*, *Myliobatis*; *d*, *Aëtobatis*.

the median or anterior ones may be elongated transversely. The fossil species, *R. Woodwardi*, has the three median vertical rows enlarged. In *Myliobatis* there is only one large median row, with three smaller ones upon either side, while in *Aëtobatis* the teeth of the median row alone remain, and are articulated to each other by a finely serrate border. These modifications are well shown in the accompanying figures. The anomalous sawfish (*Pristis*), although in no way peculiar as far as the teeth of the mouth are concerned, nevertheless possesses a remarkably elongated snout, armed upon either side by a row of hard, conical bodies usually referred to as teeth. In their histological structure they agree with true teeth, but exhibit the peculiarity of being lodged in separate sockets and growing from persistent pulps—a condition unusual among fishes. It is more than probable that they are dermal spines specially developed in this situation for some important purpose which is not at present fully determined.

I proceed next to consider the teeth of the Chimæræ (*Holocephali*), which group some authors make equal in rank with the *Elasmobranchii*, which then include the sharks and rays only. The peculiarities of the dental succession alone of this latter group, it appears to me, is quite sufficient to separate them widely from all others, and it seems somewhat remarkable that this character has never been utilized by the systematists in their schemes of classification.

The teeth of the "ratfish" (*Chimæra plumbea*) are six in number, of which two belong to the lower and four to the upper jaw. The two inferior ones may be described as broad, slightly-curved plates of moderate thickness in the form of a right-angled triangle. That border which corresponds to the perpendicular is almost straight, and is lodged in a shallow groove which runs lengthwise along the inner surface of the jaw; that which represents the base is applied to the corresponding surface of the opposite tooth; while the border representing the hypothenuse forms the free cutting edge of the tooth. This border is somewhat devious, being interrupted by three prominences. The inner surface is also slightly ribbed. The two posterior upper teeth are similar plates of a quadrilateral form with their free edges roughly serrate. The two anterior teeth above somewhat resemble ordinary mammalian incisors, and are large and scalpriform. This peculiarity has given them the name "rabbit-fish" or "ratfish." Each tooth has a cavity in the edge by which it is attached and in which the pulp is lodged. But a single set of teeth are developed during the life of the individual, and these are of persistent growth. Another living allied genus, *Callorhynchus*, is found in Australian seas, in which the teeth are similar to those of Chimæra, but in the two fossil genera, *Edaphodon* and *Passalodon*, supposed to belong to this group, the teeth are ankylosed to the jaw, which is more or less bony. On this account it is more than probable that they are to be referred to the dipnoans rather than to the chimæroids.

The Teeth in True Fishes.—The teeth of the class *Pisces*, although apparently presenting an extensive range of modification, have not, debarring the dipnoan ganoids and the plectognath teleosts, as a general rule, departed very widely from the simple conical pattern. There are some forms, however, in which the structure and arrangement are quite anomalous. It is in this group that the maximum development, as far as numbers is concerned, is reached. The salmon, pike, and some percoids may be cited in which teeth are developed in almost every conceivable part of the mouth and number many thousands; while in others, as the carps and suckers, they are few and confined to the pharyngeal bones. In others, again, as the pipefishes and sea-horses, teeth are entirely absent.

The teeth of the dipnoans are unique among fishes, and, like those of the chimæroids, are limited in number and grow from persistent pulps. The teeth of the dipnoans, the dental plates of the chimæroids, and the so-called "rostral teeth" of the sawfish are the only examples so far known of permanent teeth to be met with among piscine forms.

The dental armature of *Ceratodus Fosteri* (Fig. 194), which may be taken as illustrative of this peculiar group, has six teeth, of which four

DENTAL ANATOMY.

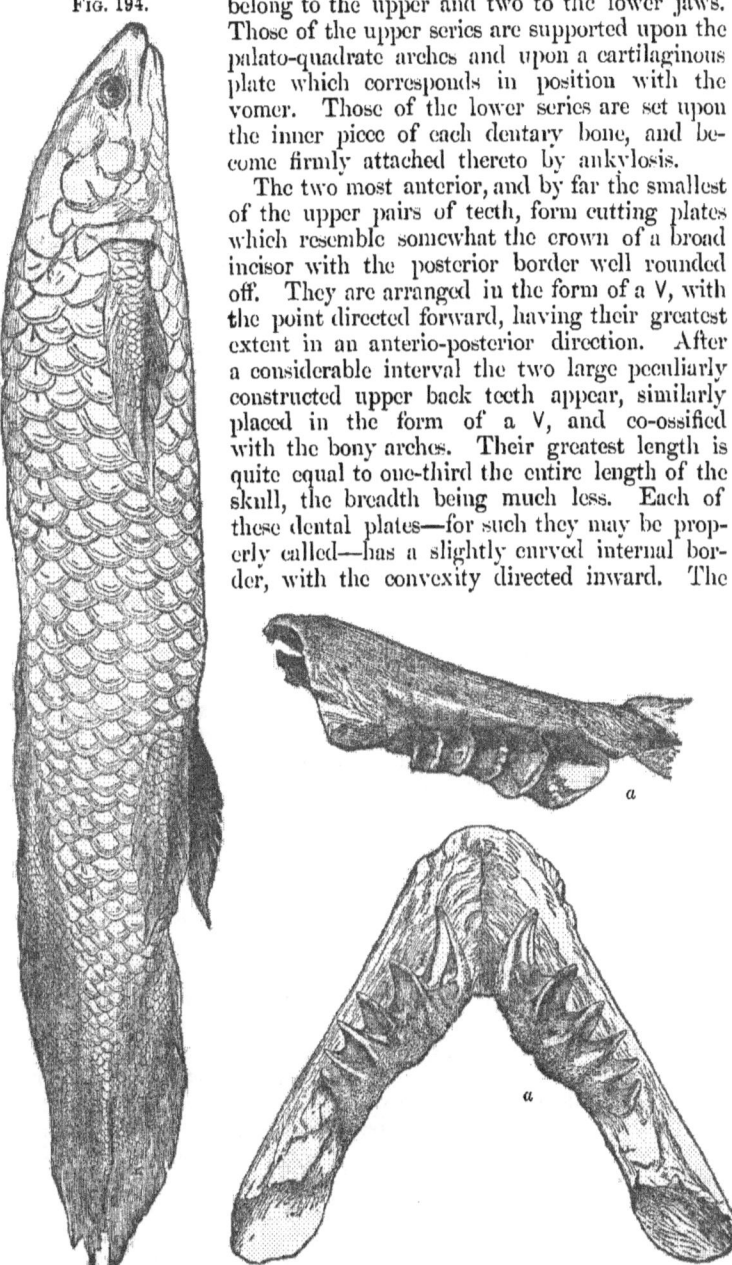

Fig. 194.

belong to the upper and two to the lower jaws. Those of the upper series are supported upon the palato-quadrate arches and upon a cartilaginous plate which corresponds in position with the vomer. Those of the lower series are set upon the inner piece of each dentary bone, and become firmly attached thereto by ankylosis.

The two most anterior, and by far the smallest of the upper pairs of teeth, form cutting plates which resemble somewhat the crown of a broad incisor with the posterior border well rounded off. They are arranged in the form of a V, with the point directed forward, having their greatest extent in an anterio-posterior direction. After a considerable interval the two large peculiarly constructed upper back teeth appear, similarly placed in the form of a V, and co-ossified with the bony arches. Their greatest length is quite equal to one-third the entire length of the skull, the breadth being much less. Each of these dental plates—for such they may be properly called—has a slightly curved internal border, with the convexity directed inward. The

Ceratodus: *a, a*, teeth of same.

lower or working face presents internally a considerable flat pitted surface reaching the entire length of the tooth, whose plane is directed outward and a little upward. It is slightly broader in front than behind. The outer border is indented by five wide vertical grooves, forming six vertically convex lamelliform projections, which encroach somewhat upon the flat surface internal to them as well-defined parallel ridges; the anterior one passes entirely across the face of the tooth, or, rather, skirts its anterior margin, and becomes continuous with the slightly elevated internal border in this situation. The grooves are deeper in front than behind, leaving the anterior projections the most pronounced, while the last one is scarcely perceptible. The points of the projections describe a gentle curve from before backward, the convexity being outward.

The two teeth of the lower jaw are very similar to those of the upper, but are somewhat narrower, owing to a decrease in width of the flat pitted surface. As already stated, they are ankylosed to two plates of bone which cover the inner and half of the lower surface of Meckel's cartilage or the central axis of the lower jaw. It will thus be seen that each ramus of the mandible consists of an inner and an outer bony plate enclosing the central cartilaginous axis, which meet in the median line below. To these is added a third piece at the symphysis on its lower surface, so that the bony part of the jaw—which in all probability corresponds with the single dentary piece in other fishes—is made up of three, one of which bears teeth. This fact is of great morphological significance, and will be referred to again when we come to discuss the attachment of the teeth.

In the two allied genera *Protopterus* and *Lepidosiren* the teeth are very similar to those of *Ceratodus*, but in numerous extinct forms referred to the dipnoans a considerable amount of variety exists. As regards the development of the teeth in this group, very little is known, and until this has been studied it will be impossible to say whether the dental plates are moulded upon a single papilla or represent the combined calcification of several. Judging from their complexity, the latter would seem to be the case.

The teeth of the remaining ganoids are of the ordinary conical form which prevails to so great an extent in the *Teleostei*. In number, position, replacement, etc. they likewise agree so closely with the average teleost dentition that it is unnecessary to make any further mention of them.

Among teleosts, however, there are several well-marked modifications in the dental armature which deserve to be noticed. One of these is presented by the *Plectognathi* (plaited jaw), of which the "trigger-fish" (*Balistes vetulus*) and the "swell toad" (*Diodon geometricus*) represent the extremes in dentition. The teeth of the mouth of *Balistes* are twenty-two in number, of which fourteen belong to the upper and eight to the lower jaw. Those of the superior series are disposed in two rows, one placed immediately behind the other, and both are lodged in the premaxillary bones. The most anterior of these rows contains eight teeth, while in the posterior one there are only six. In the front row the mesial pair are the largest, of a subtrihedral form, tapering gradually to an obtuse point. Those upon either side decrease regularly in size toward the back part of the jaw, and have notched cutting extremities.

The teeth of the posterior row are applied closely to those of the front row, and are completely concealed when the mouth is closed. They have a broader, more incisiform pattern. The teeth of the lower series are like the corresponding ones of the upper front row, and are lodged in the dentary bone. They are attached by slight ankyloses to the respective bones upon which they are supported by having their bases placed in a shallow alveolar depression, in the middle of which a conical process of bone rises up and is received into the hollow basal portion of the tooth. The successors of those in use are developed deep down in the substance of the jaws, in bony crypts which communicate with the exterior by means of foramina in the side of the jaws in the vicinity of the bases of those in use. All the teeth in this species are said by Owen to be covered with enamel. There are likewise small conical hooked teeth developed upon the pharyngeals.

While the teeth of *Balistes* are more nearly affiliated with the normal teleost condition, those of *Diodon*, on the other hand, show a much wider departure. When the mouth is closed the biting surface of the jaws seems to be invested with a continuous covering of tooth-substance; upon close inspection this is found to consist of a number of dentine plates closely incorporated with the bone of the jaws and more or less fused together at the base. Each one, however, develops separately, and takes its place when its predecessor has disappeared through wear. Just inside the margin of each jaw, in the middle line, is to be seen a broad rounded mass consisting of transverse plates of dentine intimately blended and ankylosed to the jaw bones. A faint median longitudinal suture divides each into two parts; when this becomes more distinct and extends to the edge of the jaw, as it does in some species, it constitutes the mark of generic distinction of the genus *Tetrodon*. The plates composing this mass, which is peculiar to these fishes, are developed in the same manner as the teeth, and are strictly homologous with them.

Many other examples quite as peculiar as the last one could be cited among the dental organs of fishes, but more time and space would be required than is afforded the present paper.

The mode of development is essentially the same as that described for the mammal, with the exception that the dental sacculus is generally simple; the dentine papilla arises from the corium, and the oral epithelium dips down to form a cap-like investment, in both of which calcification takes place in the manner already described. With regard to the succession and attachment of the teeth in this group, as well as in the Batrachia and Reptilia, some preliminary points in the development of the jaw bones must first be noticed.

In the early stages of embryonic development each jaw is primarily made up of two cartilaginous bars which meet in the median line in front. In the upper jaw these bars are known as the palato-pterygoid bars, and in the lower jaw as Meckel's cartilage. In the elasmobranchs these bars persist, and the teeth are supported by them. As a consequence of this condition, as we have already seen, succession in them takes place by a movement of the fibrous gum, in which the bases of the teeth are imbedded, outward over the curved surface of the jaw. Coincidently, however, with the ossification of the skeletal axis the osseous bases of the dermal denticles coalesce to form the dentary bones, as has been shown by Hertwig. By reason of the development of this bony envelope of the primitive axis of the jaw, any further movement of these denticles is prohibited, being firmly co-ossified with it.

During the coalescence of the denticles a portion of the primitive tooth-bearing membrane is enclosed beneath the fused osseous plates, and retains its original formative energy, thereby furnishing a source of supply quite equal to that of the sharks. The denticles whose basal plates form the sides and under portions of the dentary bones disappear, while those on the summit of the jaw are retained as teeth. A conformation of this position by evidence other than that afforded by embryology is seen in the dentary bones of *Ceratodus*, in which each one is made up of three or four pieces which have failed to coalesce.

The attachment of the teeth, therefore, in this group, as we would be led to anticipate, is by ankylosis to the dentary and other bones which support them. Still, there are many brush-like structures which are identical with true teeth to be found in the mouths of many fishes, which remain imbedded in the lining membrane and do not develop any connection with the underlying bones. There are several ways by which the teeth become fixed to the jaw bones in ankylosis, but the most common is for the central axis of the tooth to be occupied by a cone of osteo-dentine, which blends with the bone of the jaw. Several families of fishes have some of the teeth attached by an elastic hinge, by which they can be bent down in one direction and resume an erect attitude.[1]

TEETH OF BATRACHIA AND REPTILIA.

As we pass from the dental organs of the more typical fishes to those of frogs, salamanders, newts, etc., constituting the batrachian subdivision, a marked diminution in the number of individual teeth is to be observed. With the appearance of perfected air-breathing organs the complex hyo-branchial skeleton, typical of the fishes, becomes greatly reduced and simplified as the higher forms are approached; consequently, the branchial and pharyngeal teeth disappear in all the Vertebrata above fishes.

In all those Batrachia in which teeth exist they are usually disposed in a single row on the borders of the jaws, and are supported by the maxillary, premaxillary, and dentary bones respectively. In addition to these, each vomer (for there are two) bears a single row of teeth, between which and the maxillary row the lower jaw bites. As a rule, the Reptilia, on the other hand, lack the vomerine set, but in some of them (serpents, for example) teeth are developed upon the pterygoids and palatines, as well as upon the maxillary and dentary bones.

The teeth of the Batrachia present so limited a range of variation that the description of one will serve to give a general idea of the dentition of the whole group.

[1] This fact was first noticed by Prof. Gill.

It should be stated here that some of the tailless species (toads) are quite edentulous, while in others (the frogs) teeth are absent in the lower jaw. All the existing tailed batrachians, however, are provided with teeth which present practically the same pattern and disposition which obtains throughout the entire sub-class.

An excellent and easily-obtained example of this latter subdivision is found in the Alleghany *Menopoma*, popularly known as the "hell-bender." In this animal the skull is remarkable for its flatness and breadth, as well as the almost perfect semicircular outline which the dentigerous surface of the jaws presents. The mandible, as in the fishes, is composed of angular, articular, and dentary pieces, and is suspended to the cranium by means of two bones known as the squamosals through the intervention of the quadrates.

The palato-quadrate or palato-pterygoid arch is not so well defined, although the principal elements are present. The vomers are two in number, and occupy their usual position behind the maxillaries, sharing in the formation of the bony roof of the mouth. The maxillaries and premaxillaries also have the same position as in the fishes, but are less mobile, on account of sutural connections with the surrounding bones.

The biting surface of each jaw is produced into a sharp ridge by reason of the existence of a well-marked ledge extending the full length of its internal face. This ledge is converted into a groove in the recent state by a fold or flap of the gum, which forms its internal wall, and is in all probability homologous with a similar structure (the thecal fold) found in the sharks. At the bottom of this groove the tooth-germs of the successive sets of teeth are developed. It will be seen, therefore, that the general arrangement is not different from that of the sharks; but this important difference is to be observed: in the sharks the bases of the teeth are at first directed upward, and it is only when they are about ready to take position on the working surface of the jaw that they assume the erect attitude; this, as we have already seen, is due to the movement of the entire gum outward. This manner of replacement gives weight to the conclusion that the teeth in the sharks are invaginated dermal spines, the position of which we would expect to find reversed upon the inside of the jaws.

In the batrachians, on the contrary, the teeth are said to have an erect position from the earliest stages of development, and it is less easy to see how they represent dermal spines or how the position came to be reversed. Believing, however, that all the maxillary and mandibular teeth were originally of tegumentary origin, as is clearly demonstrable in the sharks, it is more than probable that the arrest of the outward movement of the gum in the batrachian by the appearance of ossifications around Meckel's cartilage to form the dentary bones is responsible for this change. It is quite possible that the tooth-germs of the batrachian do at first have the same position as those of sharks—that is to say, with the points directed downward—and that the formative energy of the tissues beneath causes them to become erect at a comparatively early period.

We have already stated, in connection with the account of the dental organs of the elasmobranchs, that tooth-succession is primarily due to the evagination of the lining membrane of the oral cavity. Whether this is caused by unequal rate of growth of the surrounding parts, whereby this lining membrane is forcibly pulled outward and replaced from within, or whether the formative energy inherent in the membrane itself causes it, is difficult to determine. This primary cause of succession is profoundly affected by the development of an osseous sheath from this same membrane around the central cartilaginous axis of the jaws. It seems plausible that secondarily the cause of tooth-succession is to be sought for in the proliferation of the cellular elements beneath the young and growing germ.

To the inner side of the external osseous wall of this groove in *Menopoma* the functional teeth are attached; their bases are slightly enlarged and extend quite to the bottom of the groove, while the tapering crowns reach considerably above the level of the jaw. Attachment takes place by the ankylosis of that part of the base which is in contact with the outer wall to the bone of the jaw through the intermediation of osteodentine. This manner of implantation is known as "pleurodont," on account of the fancied resemblance of the teeth so attached to ribs.

The teeth in use at any given time are from thirty to forty in number upon either side in each jaw; they are subequal in size, and are placed with great regularity, being separated by spaces about equal to the width of a single tooth. Their crowns are sharp-pointed and slightly recurved; they are said in some species to be tipped with enamel, which is probably true of all. The vomerine teeth are fewer in number than the maxillary, there being not more than twelve or fifteen upon either side. Their line of direction and manner of implantation coincide with the maxillary row external to them, agreeing with those also as to size and form.

In some of the extinct batrachians, notably the labyrinthodonts, there were several teeth of the maxillary and premaxillary set considerably enlarged and of a caniniform pattern. The species of this section were mostly of large size and presented a formidable dental armature. They likewise differ from all other of the Batrachia in that the teeth were implanted in distinct sockets, and were rarely if ever attached to the body of the jaw by ankylosis. The structure of the teeth is a curiously complex one, and finds no parallel throughout the entire Vertebrata, save in one extinct saurian (*Ichthyosaurus*) and several fishes, which exhibit a similar condition in a less perfected degree.

The external surface of the crowns of the teeth in the labyrinthodonts is marked by a number of longitudinal ridges, separated by what at first sight would appear to be comparatively shallow grooves extending from the base to the apex of the crown. Upon cross-section, however, these fissures are seen to penetrate into the body of the tooth to a remarkable depth—to a point, in fact, quite near the pulp-cavity or central axis of the tooth, where they are separated from it by a thin wall of dentine. The entire outer surface is covered by a thin layer of cement, which is reflected inward to the bottoms or internal terminations of the fissures just mentioned. The cut edges of this reflected layer of cement, which is of uniform thickness throughout, are almost straight for a short distance beneath the surface, but soon become very tortuous. The arrangement of the dentine is as follows: The axial portion of the tooth consists of a central cone of dentine hollowed out in the centre to receive the pulp. In cross-section this cone appears as a ring surrounding the pulp-cavity; from it plates of dentine, which are cleft by the fissures from without, radiate to the periphery, pursuing the same tortuous course as that of the fissures. These dentinal plates are separated from each other by fissures which radiate from the axial cavity, but do not reach the exterior of the crown. Some of the dentinal plates do not arise from the central ring, but appear on transverse section as processes from the periphery of the crown directed toward the central axis, thus causing the fissures which radiate from the pulp-cavity to become bifurcated at their outer or peripheral terminations. Some of these accessory processes reach but a short distance toward the interior, while others penetrate halfway or more to the centre.

This complexity in the arrangement of the tooth-substances has suggested the name of the typical genus, *Labyrinthodon*, which was originally described by the great anatomist Prof. Owen. Just how it has been produced is difficult to understand.

TEETH OF THE REPTILIA.—This class of vertebrated animals includes snakes, lizards, crocodiles, turtles, etc., and, considering the extinct as well as the recent forms, is divisible into eleven distinct orders, according to Prof. Cope's classification. Of these, but five are represented in the existing fauna, the others having become extinct in the different epochs of the earth's history.

The batrachians make the nearest approach to the permanent gill-breathing vertebrates in all the essential features of their structure; the Reptilia, on the other hand, furnish us with the transitional forms leading to the avian and mammalian stems. It is a very significant fact—and one upon which the doctrine of evolution is primarily based, so far at least as the Vertebrata are concerned—that the lowest forms appeared first in the order of time, and were followed by those higher in the organic scale; thus we have the cartilaginous fishes as the earliest representatives of vertebrated animals; after them come the batrachians; next the reptiles and birds, and finally mammals. It must be borne in mind, however, that the highest of one group is not always most nearly related to the group next above it; for example, if we compare the structure of a bony fish with that of a salamander, a great interval will be found to exist, but if we institute a comparison between the latter and a dipnoan fish, which is comparatively little removed from the cartilaginous forms, this interval will be found to be materially diminished. Thus, the conclusion is obvious that the Batrachia sprang not from the higher bony fish, but from some generalized representative of the piscine type. The same reasoning can be applied to other divisions.

As regards the Reptilia, the chief distinction between them and the Batrachia consists in the circumstance that the latter during the larval stages of their existence breathe by means of gills like the fishes, whereas the Reptilia breathe by means of true lungs from the time of birth. Important osteological differences are found in the bones of the skull.

The earliest appearance of the Reptilia dates back to the Permian Epoch, where they are represented by a group of peculiar batrachian-like reptiles which has been designated the *Theromorpha* by Prof. Cope. This group includes two important divisions, one of which, the *Anomodontia*, was first described by Prof. Owen from the Triassic (?) deposits of South Africa; the other is the *Pelycosauria*, which is so far known only from the American Permian.

The osteological structure of this order furnishes us many transitional characters between the Batrachia and more typical Reptilia, on the one hand, while on the other they seem to stand in ancestral relationship to the prototherian Mammalia. Their batrachian affinities are manifested in the structure of the pectoral and pelvic arches, in the structure of the limbs, and the possession of teeth on the vomer. In the absence of a parasphenoid bone in the base of the cranium and the unicondylian condition of the skull they are markedly reptilian. The structure of the pelvic and pectoral arches and limbs, together with the intercentral articulation of the ribs, allies them with the lower Mammalia.

Their dental organs present a considerable variety of structure—in some instances departing widely from the simple conical form usual among the other orders of this class of the Vertebrata. In one genus (*Dimetrodon*) there were two large caniniform teeth in each premaxillary, implanted in distinct sockets; these were followed by a single row of maxillary teeth, whose crowns resemble somewhat the premolars of the dog in general pattern. They were lodged in distinct alveoli, and exhibit the remarkable peculiarity of being implanted by double fangs, or rather single ones deeply grooved upon either side, otherwise unknown among the Reptilia. The first tooth behind the maxillo-premaxillary suture is enlarged into a canine, and the entire maxillary series does not exceed fifteen in number. The palato-pterygoid arch is present, and one of its elements, probably the pterygoid, is thickly studded with small conical teeth irregularly disposed. Another element which lies internal to this last-mentioned bone is described by Prof. Cope as bearing a single row of teeth. Other genera related to this one are *Theropleura*, *Clepsydrops*, etc. of Cope, which present minor differences in the form and size of the corresponding teeth.

A nearly-allied family of this group is the *Diadectidæ*, likewise described by Cope. A typical example of the dentition of this family is seen in *Empedocles molaris*, wherein the pattern of the crowns of the molars is thoroughly unique. The teeth are disposed on the borders of the premaxillary, maxillary, and dentary bones, as well as upon the vomer, which forms a median keel in the roof of the mouth. The maxillo-premaxillary set in the upper jaw describe a sigmoid curve in their line of implantation, and form an uninterrupted series from the front to the back of the mouth. There are fourteen teeth belonging to this series, of which the first two are larger than those immediately succeeding them. They have obtuse subconic crowns, and are lodged in distinct alveoli. From this point the teeth gradually decrease in size up to the sixth, when they again become larger and more complex in pattern. The crowns of the typical molars have a much greater transverse than longitudinal extent; the grinding surface is somewhat elliptical in outline, and is provided with a submedian cusp which stands nearer the outer than the inner border. The portion of the crown external and internal to the median cusp is horizontal, and has its surface thrown into conspicuous folds or wrinkles. The teeth of the lower jaw are essentially like those of the upper. The vomerine teeth are small and conical, and are disposed in two longitudinal rows.

In the typical genus (*Diadectes*) there is a well-developed canine, while in another member of the family (*Helodectes*) there are canines and a double row of maxillary teeth upon either side in the dentigerous surface of these bones.

In the other subdivision of this order—viz. *Anomodontia*—the dentition is reduced to large pointed, recurved tusks, which are lodged by distinct sockets in the maxillary bones. The rest of the jaw is edentulous, and was in all probability ensheathed in a corneous substance, as in the existing turtles. Other extinct members of this subdivision, notably *Rhynchosaurus* and *Oudenodon*, were entirely edentulous, and in all probability were the ancestors of the turtles.

That division known as the *Crocodilia* includes the alligators, crocodiles, gavials, etc., which are separated from the other Reptilia by a number of important osteological characters, prominent among which is the complete development of the bony roof of the mouth. Teeth are supported by the premaxillary, maxillary, and dentary bones only, the palatines and pterygoids having approximately the same relations and edentulous condition as in the mammalian skull. In no crocodilian so far known are the teeth ever ankylosed to the body of the bones upon which they are borne, but, on the contrary, they are set in distinct sockets disposed in a single row along the margins of the tooth-bearing bones. In young specimens the alveoli are apt to be ill defined, more especially toward the back part of the jaws, but as age advances the bony partitions become more distinct. On account of each tooth having a distinct alveolus, this division of the Reptilia was formerly known as the thecodonts, in contradistinction to the pleurodonts—a condition already mentioned in connection with *Menopoma*—and acrodonts, presently to be described.

A good example of the dentition of a crocodilian reptile is afforded by the Mississippi alligator (*Alligator Mississippiensis*), which can be found in almost any osteological collection in this country. In the upper jaw there are from eighteen to twenty-two teeth upon either side, of which five are usually set in each premaxillary and the remainder in the maxillary bones. The most anterior of the premaxillary series is the smallest, from which they gradually increase in size to the fourth, which is nearly twice as large as any of the others; the fifth is about equal to the third. The first of the maxillary series is likewise the smallest; the three succeeding teeth gradually increase in size until the third is reached (the ninth counting from the first tooth in the premaxillary), which is known as the canine of the upper jaw. The eighth and tenth are frequently as large as the canine. Behind, the teeth become smaller, and are again enlarged in the vicinity of the sixteenth or seventeenth from the first premaxillary tooth; from this point they rapidly diminish toward the posterior end of the tooth-line.

In the lower jaw the teeth are likewise of unequal proportion, but those which are largest in the one series are opposed by the smallest of the opposite set; thus that tooth which is caniniform in the lower jaw is the fourth, and bites in front of the corresponding tooth above. It is received into a deep fossa in the upper jaw just internal to the alveolar border at the point of junction of the maxillary with the premaxillary bone, or between the fifth and sixth teeth above. It not unfrequently happens in old specimens that this fossa is converted into a foramen leading to the external surface of the skull by the perforation of its base. In such cases the point of the lower canine passes through the upper jaw and appears upon the upper surface.

The only important distinction between the alligators and the crocodiles consists in the fact that in the latter this fossa is open externally, causing the tooth-line to be interrupted by a deep notch, whereas in the latter it is intact.

Both the alligators and the more typical crocodilians are remarkable for the breadth of the palate and the flatness of the muzzle, as well as the alternate increase and decrease in the size of the teeth from before backward; but in the gavials the snout is very long, narrow, and almost cylindrical; the teeth, too, are more nearly equal and of more regular proportions.

In the alligator the anterior teeth have conical crowns terminating in sharp points, which are slightly recurved. The posterior ones have more obtuse crowns, which terminate below by a moderately well-defined neck. In some species the anterior and posterior surfaces of the crowns are produced into trenchant edges, which may be more or less serrated; in the alligator this is but faintly marked.

The manner of succession is not different from that of the other lower vertebrates. If the root of a tooth in place be exposed, the successional sets in various stages of development will be seen below and to the inside of it, arranged in the form of a nest of crucibles. This arrangement results by reason of the absorption of the inner wall of the root of the tooth in place which the immediate successor causes. By this means the point of its crown comes to occupy the pulp-cavity of the functional tooth.

In the order *Lacertilia*, which includes the lizards proper, a more varied development of the dental organs is met with. As a general rule, teeth are borne upon the pterygoid and palatine as well as upon the maxillary, premaxillary, and mandibular bones. There are, however, some exceptions, one of which is afforded by our little "horned toads" (*Phrynosoma*), in which the palatines and pterygoids are edentulous. The teeth may be either "pleurodont" or "acrodont" in their manner of implantation, but in certain extinct forms (*e. g. Mososaurus*) both conditions are to be observed. In the case of acrodontism the bases of the teeth are soldered to the summits of slight elevations which arise from the alveolar border of the jaws. Pleurodontism, as has already been mentioned, consists in the ankylosis of the base and outer sides of the teeth to the outer wall and bottom of the dental groove. Another variety of implantation, known as ccelodontism, has been described, in which the tooth has a permanent pulp-cavity, and is attached to the outer wall, leaving the base free; it should be mentioned that in pleurodonts the pulp-cavity is not permanent; it soon becomes obliterated, leaving the tooth solid.

A fair example of a pleurodont lacertilian is afforded by the majority of the numerous species of the *Iguanidæ*, although some of the members of the iguanian family, such as *Isturus*, *Lophyrus*, *Calotes*, and others, are acrodont. In the horned iguana (*Metopocerus cornutus*) the maxillary and premaxillary teeth are from twenty-two to twenty-three in number upon either side. The central ones of the premaxillary set, of which there are four, are smallest, the outer ones slightly enlarged. These, together with the first five or six maxillary teeth, have subconic recurved crowns, while the crowns of the posterior maxillary series are laterally compressed into anterior and posterior cutting edges and terminated by a principal cusp. Of the two edges, the anterior is the longer and is interrupted by three minor cusps, the posterior being shorter and bearing only a single accessory cusp. The presence of these cusps gives the crown a serrated appearance when viewed from the side.

The teeth of the lower jaw are from twenty to twenty-two in number upon either side, and are similar in form to those above, with the exception that there are generally two accessory cusps upon either trenchant edge of the crown. There is in addition to these a single row of small conical teeth supported by each pterygoid bone; the number of these varies from five to seven.

The only lacertilian which is known to be poisonous is the "Gila monster" (*Heloderma suspectum*) of our American fauna. Recent experiments of Drs. Mitchell and Reichart of Philadelphia have demonstrated beyond doubt the poisonous qualities of its salivary secretion. Considerable interest therefore attaches to its dental organs, as well as to the anatomy of the poison-glands; this latter subject I am, unfortunately, not in a position to describe, and will therefore limit what I have to say here to a consideration of the teeth only.

This animal, of which there are two species, is confined to the desert wastes of the South-western United States, where it is not of rare occurrence. In life it has a rather repugnant appearance, which is no doubt increased by our knowledge of its poisonous qualities. It attains a length of eighteen inches or two feet, and is covered with bright yellow spots, a circumstance which gives the name *Heloderma* to the genus, meaning "sun skin." Its venomous nature was not known until the experiments above mentioned were made, although Prof. Cope had reason to suspect as much, and gave the name "suspectum" to the species, which he described several years before.

The teeth are supported by the premaxillary, maxillary, and dentary bones, the palatine and pterygoids being edentulous. Those of the premaxillary, of which there are three upon each side, are the smallest of the upper teeth. They increase regularly in size from before backward, and form a continuous series, with the maxillary teeth behind, which continue to augment their dimensions up to the eighth tooth from the median premaxillary pair or the fifth of the maxillary set. From this point backward the two remaining teeth become slightly smaller. The teeth of the lower jaw are nine in number, and are disposed very much in the same manner as those above—the smallest in front and the largest toward the back part of the mouth. A considerable disparity in size exists between the inferior series and the corresponding teeth above, those below being much the longer and more robust.

In their manner of implantation they cannot be said to be either acrodont or pleurodont, but rather intermediate between the two. The internal aspect of each jaw, which is remarkable for its breadth, is slightly bevelled internally, causing the outer edge to rise a little above the inner. Nearer the outer than the inner edge of this bevelled surface are a number of low bony elevations, corresponding to the number of the teeth in functional use, to the summits of which they are attached by ankylosis. In some instances these elevations are so faintly indicated that the teeth appear to be soldered to the bevelled surface of the jaw directly. Just internal to the basis of the functional teeth may be seen the successive sets in different stages of development. In the recent state they are covered by a fold of the gum, which likewise covers up the bases of the functional teeth.

The form of the crown is that of a long, slender, sharp-pointed cone curved inward and backward. The anterior surface of each tooth is marked by a well-defined groove extending from the base to the apex. It is somewhat deeper at the base than the summit, and is most distinct in the teeth of the lower jaw. The intervals between the bases of the teeth allow abundant room for the accommodation of poison-glands, the secretion of which is conveyed down these grooves and thus injected into the wound which the teeth inflict upon a prey.

Another group of curious and interesting reptiles is the *Dinosauria*, which became extinct at the close of the Cretaceous Epoch. They are of especial interest on account of their remarkable bird-like affinities, and, according to the views of many authors, were the direct progenitors of the struthious birds, or ostriches, emus, etc. They were mostly of gigantic size, and some of them are remarkable for the great number of teeth contained in the upper and lower jaws; others, again, were almost edentulous.

In the iguanodonts and hadrosaurs, which are typical representatives of the herbivorous division of this order, the crowns of the teeth are somewhat expanded and are marked externally by vertical ridges, while the internal portion is smooth and rounded. In *Iguanodon* the external surface, to which the enamel is confined, is traversed by three vertical ridges, separated by vertical grooves; the anterior and posterior edges were serrated, as in *Iguana*, before the crown was abraded by wear. In the hadrosaurs there is but one vertical ridge, which is external in the upper and internal in the lower teeth. The part which bears this ridge is known as the enamel or cementum plate. Prof. Cope has recently had the opportunity of satisfactorily determining the dental peculiarities of this group of gigantic saurians, as exemplified by the genus *Diclonius*, through the fortunate discovery of an almost complete skeleton by Dr. Russel Hill and the author in the Bad Lands of Dakota during the summer of 1882.

According to Prof. Cope's description, there are in all two thousand and seventy-two teeth. Of these, there were not more than two or three hundred in use at one time, the others being arranged in successive rows beneath, ready to take the place of the functional ones when they were worn out. One striking peculiarity which this reptile presents is in the dentigerous character of the *splenial* and the edentulous condition of the dentary bones of the mandible. The teeth are relatively small, and are placed at some distance from the anterior part of the mouth. This part of the jaws is believed to have been occupied by a kind of horny sheath similar to that found in birds and turtles.

The proportions of the limbs were those of the kangaroo, the posterior greatly exceeding the anterior in size. The general shape of the skull is very much like that of a bird with a large spatulate beak; it was supported upon a long, flexible neck, which was doubtless useful to the animal in gathering the soft aquatic vegetation upon which, from the character of its teeth, it is supposed to have subsisted. It likewise had a powerful tail, much deeper than thick, which probably served not only as a fifth limb in balancing the weight of the animal, but could also have been useful as a swimming organ. The feet were provided with true hoofs.

The carnivorous dinosaurs were scarcely inferior in size to the herbivorous species, but were of a more slender and active build. Their jaws were provided with large, powerful conical teeth, better adapted for the capture of living animal prey. The terminal phalanges were ensheathed in distinct claws.

Another order of the Reptilia, and one which is probably best known, is the *Ophidia*, or snakes. Especial interest attaches itself to the dental organs of many of this group, inasmuch as their poisonous bite constitutes one of their most conspicuous features and renders them particularly obnoxious as well as dangerous to life.

According to most systematists, the order is divisible into five sub-orders, which have been defined as follows:

I. "The palatine bones widely separated, and their long axes longitudinal; a transverse (ecto-pterygoid) bone; the pterygoids unite with the quadrate bones."
  a. "None of the maxillary teeth grooved or canaliculated" . . . . . . *Asinea.*
  b. "Some of the posterior maxillary teeth grooved" . . . . *Tortricina.*
  c. "Grooved anterior maxillary teeth succeeded by solid teeth" . *Proteroglyphia.*
  d. "Maxillary teeth few, canaliculated, and fang-like" . . . . . *Solenoglyphia.*
II. "The palatine bones meet or nearly meet in the base of the skull, and their long axes are transverse. No ecto-pterygoid bone; the pterygoids are not connected with the quadrate bones" (Huxley) . . . . *Scalecophidia.*

The first of these sub-orders includes nearly all of the harmless or non-venomous species, of which the black snake, garter snake, boa, etc. are familiar examples. The second includes a single family with few species, said to be harmless; they are confined to Africa. The third sub-order embraces such forms as the deadly cobra, the coral snake, harlequin snake, and others. The fourth includes the vipers, rattlesnakes, adders, etc. The last is represented by few species which are non-venomous.

In general, the dentigerous elements of the ophidian skull may be said to consist of maxillary, palatine, and pterygoid bones of the upper and the dentary bones of the lower jaw, although in the pythons and tortrices teeth exist upon the premaxillaries as well. In *Rachiodon*, a singular African species of the Asinea, the teeth of the jaws are extremely small and soon disappear. This loss is compensated for by an excessive development of the hypopophyses of several of the anterior vertebræ, which pierce the superior wall of the œsophagus and are tipped with a layer of hard cementum. The food of this species consists of the eggs of small birds, which it swallows whole. During the act of deglutition the calcareous shell is brought into contact with and crushed by these œsophageal teeth, thus preventing the escape of any of the nutritious substances.

In the non-venomous species the maxillary bone is long, and bears a row of teeth which are of variable size in the different parts of the jaw in different genera. In some the teeth are largest in front and smallest behind; in others it is the reverse of this; while many have the teeth of equal size throughout; then, again, certain teeth of either jaw may be specially enlarged and separated from the others by a diastema. All these conditions have received distinct names.

All serpents are acrodont, and the crowns of the teeth consist of long, sharp-pointed, recurved cones which are designed more to prevent the escape of a struggling prey than as instruments of mastication. The two rami of the lower jaw are bound together at the symphysis by elastic ligaments, which, together with the great distensibility of the throat, due to the mobility of the suspensory bones, allows them to swallow objects many times larger than the usual diameter of the body. During the act of swallowing the recurved and pointed teeth act as so many hooks to prevent a backward movement of the object.

In the sub-order known as the *Proteroglyphia* the maxillary bone is shortened somewhat, and the anterior teeth are enlarged and grooved on their anterior faces. One of these teeth (the anterior) is the largest, and is denominated the fang. It is permanently erect in these serpents, being ankylosed to the maxillary bone, which is capable of comparatively little movement.

In the solenoglyphs, on the other hand, of which the rattlesnake is an excellent example, the maxillary bone attains its maximum of abbreviation and supports a single tooth, the fang.[1] It is movably articulated with the lachrymal above by means of a ginglymoid joint. The fang is canaliculated or perforated in the direction of its long axis by a canal which opens near its point. This canal results from the fusion of the free edges of the anterior groove, which remains open in the fangs of the proteroglyphs. When the mouth is closed, the maxillary bones are retracted and the fangs lie parallel with the roof of the mouth; when the animal "strikes," the maxillary bones are extended by special muscles and the fangs become erect.

The canal of the fang receives at its proximal termination the duct of the large poison-gland, which lies above it, so that when the punctured wound is inflicted the poisonous secretion is injected into it. This is facilitated by a coincident contraction of the muscles which surround the gland. It has been suggested by Owen that as the quantity of saliva and lachrymal secretion is increased during particular emotions, so the rage which stimulates the venom-serpent to use its deadly weapon must

[1] Usually, a number of teeth are found just behind the fang in this bone, some of which are nearly or quite as large as the fang itself. These are the teeth which are destined to succeed the functional fang whenever it shall have been shed or lost by accident.

be accompanied with an increased secretion and great distension of the poison-glands.

In reference to the poisonous character of this secretion, it is a well-known fact that the normal saliva of many animals is more or less dangerous when injected directly into the blood, and that in a state of rage it is rendered more so. Prof. Cope has recently called my attention to the possible explanation of the poisonous character of this analogous secretion of the venomous serpents: that since their peculiar method of locomotion would expose them most frequently to injuries and inconveniences calculated to excite this state, the normal salivary secretions have become accordingly modified.

The remaining orders of the Reptilia do not exhibit any important modifications of the dental system worthy of special notice.

THE TEETH OF THE MAMMALIA.

WITH a consideration of the teeth of the Mammalia we enter upon a study of a series of dental organs whose complexity, variety, and specialization surpass those of any other group of the Vertebrata. The wide diversity of conditions under which the different members of this great group exist would of itself lead one to anticipate a corresponding diversity in dietetic habits, as well as organs suitable for the prehension and assimilation of the substances by which they are nourished. The broad grinding surface afforded by the molar tooth of the elephant, the sharp, trenchant, sectorial dentition of the lion, the great scalpriform incisors of the beaver, the small cylindrical teeth of the armadillo, are a few examples of the great range of variety which mammals exhibit in the form of their dental organs.

As already remarked in the introductory pages, this study is greatly facilitated by considering it from the standpoint of evolution, or rather in the light of the palæontological history of the group. If we look upon the fossil remains of any given period of geologic time as the representatives in part of the animals which at that time inhabited the earth, it then becomes of the utmost importance to ascertain the exact relationship which the animals of each period bear to those which have preceded and succeeded them in time. It is needless to say that the conclusions which we are compelled to draw from studies of this character are important and significant, and serve to bring into the closest connection many isolated facts which if considered by themselves would be wholly unintelligible.

Some objection to this method of treatment will doubtless be raised by those who do not accept evolution as a demonstrated fact, or those, again, who consider our information concerning extinct forms too meagre for purposes of generalization. In answer to these objections it must be urged that palæontological law compels us to recognize the important fact that in every department of life the generalized has preceded the specialized in time; we pass from the simple to the complex, whether an individual organ or the entire organism be considered; and the teeth form no exception to this rule. So conclusive is the testimony which it is now possible to adduce in support of this general proposition, and so pregnant are the minds of modern biologists with this belief, that it seems utterly impossible to escape the conviction that life from its earliest inception has been continuously, and in many instances progressively, modified. As to the nature of the causes which have induced this modification, there is much less unanimity of opinion. It is a question regarding which the most exhaustive philosophic discussion is now in progress.

When we speak of the origin of mammalian teeth, it is necessary to have some definite knowledge of the origin of this class of animals before we can be absolutely certain of just what constitutes a primitive mammalian dentition. Unfortunately, the evidence which would enable us to determine the ancestry of the mammal beyond dispute has not as yet been found, but it appears sufficiently evident that we are limited in our choice to the Batrachia and Reptilia of the Permian Period. Huxley, who has devoted considerable attention to this subject, concludes that we must go backward past the Reptilia directly to the Batrachia. This conclusion is primarily based upon a comparison of the pelvic arch of the monotremes with that of the batrachians. In addition to the evidence drawn from this source, upon which his argument is principally founded, the following reasons are given for this view: "The Batrachia are the only air-breathing Vertebrata which, like the Mammalia, have a dicondylian skull. It is only in them that the articular elements of the mandibular arch remain cartilaginous, while the quadrate articulation remains small, and the squamosal extends down over the osseous elements of the mandible, thus affording an easy transition to the mammalian condition of those parts. The pectoral arch of the monotremes is as much batrachian as it is reptilian or avian. The carpus and tarsus of all Reptilia and Aves, except the turtles, are modified away from the batrachian type, while those of the mammal are directly reducible to it. Finally, the fact that in all Reptilia and Aves it is a right aortic arch which is the main conduit of arterial blood leaving the heart, while in the Mammalia it is the left which performs this office, is a great stumbling-block in the way of the derivation of the Mammalia from any of the Reptilia or Aves. But if we suppose the earliest forms of both Reptilia and Mammalia to have had a common batrachian origin, then there is no difficulty in the supposition that from the first it was the left aortic arch in the one series, and the right aortic arch in the other, which became the predominant feeder of the arterial system."

If we had only the recent forms to consider, the argument advanced by this learned anatomist would be specially potent; but when we study carefully the osteology of the Reptilia of the Permian Period, many of the arguments here advanced are invalidated. The structure of the pectoral and pelvic arches of the theromorph Reptilia, as has been ascertained by Cope, resembles that of the monotremes far more than does that of any known batrachian. The carpus and tarsus of these forms are almost identical with those of the monotremes, while comparatively little importance can be attached to the dicondylian character of the skull, from the fact that there is in certain members of this group a double articular surface on the occipital bone for the atlas vertebra. The only osteological character left in which the Batrachia resemble the Mammalia most is that of the quadrate articulation; which resemblance is somewhat counterbalanced by the approaches to the distinctive peculiarities of the mammalian dentition found only in the *Theromorpha*. The condition of the arterial system must remain inferential for this group, since it became extinct, so far as we now know, at the close of the Permian Period. Upon the whole, I am disposed to think that there are quite as many reasons to regard the theromorph Reptilia as the ancestors of the mammal as there are to regard in the same light any of the Batrachia so far discovered.

Accepting the "placoid scale" or the "dermal denticle" as the structure from which all teeth were primarily derived, we have, as characters of a primitive dentition, the following: (1) the conical form; (2) increased number; (3) frequent and almost endless succession. These conditions we have fulfilled in many of the sharks. The next step in specialization consists in the fusion of the basal osseous plates of the "dermal denticles" to form the maxillary and dentary bones, to which the teeth become attached by ankylosis. This, we have already seen, obtains in a majority of the fishes, and is associated largely with the simple conical form. In the Batrachia the conical form, this mode of attachment, as well as the succession, are closely adhered to, but the individual teeth are reduced in number. In certain of the Reptilia—*e. g. Theromorpha*—another advance is made in the implantation of the teeth in distinct sockets, with a disposition to form more than one root or fang. There are still, however, many successive sets of teeth developed. Lastly, in the Mammalia the teeth are generally greatly reduced in number; they are always implanted by one or more roots in a distinct socket, and there are never more than two sets developed, the second of which is only partially complete; they are also, as a general rule, of a complex nature and show a wide departure from the primitive cone.

In searching, therefore, for a primitive or generalized mammalian dentition, the most important point to be taken into consideration is the following: numerous single-rooted teeth, confined to the maxillary and mandibular bones, implanted in distinct sockets, with a complete development of one or more successive sets. It is possible, even probable, that this stage in tooth-development was reached in the ancestors of the Mammalia before they assumed their distinctive characteristics as such; but the nearer any approach is made to this condition on the part of the mammal, in that proportion it may be regarded as primitive in its dental organization.

Having already spoken of the probable origin of the Mammalia, it now remains to give a brief synopsis of their classification before proceeding to a detailed description of their teeth. The arrangement here adopted is, with some modification, the one which has been proposed by Prof. E. D. Cope, and is based upon a study of both fossil and recent forms:

TEETH OF THE VERTEBRATA. 393

TABLE OF CLASSIFICATION OF MAMMALIA.

MAMMALIA
- PROTOTHERIA — Monotremata (duckbill).
- EUTHERIA
  - Didelphia — Mutilate Series
    - Polyprotodontia (opossum).
    - Diprodontia (kangaroo).
    - Plagiaulacidæ (extinct).
  - Monodelphia
    - Ungulate Series
      - Sirenia (sea-cow).
      - Cetacea
        - Zeuglodontia (extinct).
        - Denticete (toothed whales).
        - Mysticete (whalebone whales).
      - Bunotheria
        - Insectivora (shrew, etc.).
        - Tillodontia (extinct).
        - Toxodontia (extinct).
        - Prosimiæ (lemurs).
      - Rodentia
        - Sciuromorpha (squirrels).
        - Myomorpha (rats).
        - Hystricomorpha (porcupines).
        - Lagomorpha (rabbits).
      - Cheiroptera (bats).
      - Carnivora
        - Fissipedia (dogs, cats, etc.).
        - Pinnipedia (seals).
      - Primates (monkeys, man, etc.).
    - Ungulate Series
      - Taxeopoda
        - Condylarthra (extinct).
        - Hyracoidea (hyrax).
      - Amblypoda (extinct).
      - Proboscidea (elephants).
      - Diplarthra
        - Artiodactyla (hog, cow, deer, etc.).
        - Perissodactyla (horse, tapir, etc.).

It will be seen, from the foregoing table, that the Mammalia are divisible into two primary groups, which hold the rank of sub-classes. The first of these, *Prototheria*,[1] has but two living representatives, both of which are confined to the continent of Australia. These are the *Echidna*, or spiny ant-eater, and the duck-billed platypus. The principal characters by which they are separated from all other Mammalia may be conveniently contrasted with those of the second sub-class, *Eutheria*, as follows : in the former there are (1) " large and distinct coracoid bones, which articulate with the sternum. (2) The ureters and the genital ducts open into a cloaca into which the urinary bladder has a separate opening. (3) The penis is traversed by a urethral canal which opens into the cloaca posteriorly, and is not continuous with the cystic urethra. (4) There is no vagina. (5) The mammary glands have no teats." In the *Eutheria*, on the other hand, (1) "the coracoid bones are mere processes on the scapula in the adult, and do not articulate with the sternum. (2) The ureters open into the bladder, the genital ducts into a urethra or vagina. (3) The cystic urethra is continuous with the urethral canal of the penis. (4) There is a single or double vagina. (5) The mammary glands have teats." (Huxley).

In their anatomical structure the *Prototheria* resemble the reptiles and birds more than does any other mammal. This is particularly conspicuous in the pectoral arch and the reproductive system. On this account, De Blainville applied the name *Ornithodelphia* (bird womb) to them, by which they are sometimes known. Strange as it may seem, no fossil remains of great antiquity of this most primitive group of all Mammalia are with certainty known to exist, but it may yet be found that the earliest mammalian representatives, which date as far back as the Triassic Period, and which are known from teeth and jaw bones only, really belong to the *Prototheria* rather than to the *Didelphia* or pouched series of the *Eutheria*, as is frequently maintained. Both the living forms are devoid of true teeth.

The second sub-class, *Eutheria*, has two principal divisions : *Didelphia* (double womb), including those animals popularly known as the "pouched quadrupeds," of which the opossum, kangaroo, wombat, etc. are familiar examples; and the *Monodelphia* (single womb), which embraces all the remaining mammals. The name of the first subdivision, *Didelphia*, was applied by its author, De Blainville, with reference to the peculiar habit which these animals possess of sheltering their helpless young in an abdominal integumentary fold. This is correlated with the only important character in which they differ from the monodelph division—viz. the complete absence of an allantoic placenta or any uterine connection between mother and fœtus. In consequence of this peculiarity of gestation the young are born in an exceedingly helpless and imperfect condition, and are nourished for a considerable period in the marsupium or pouch of the mother. This character is

---

[1] The classification of the Mammalia proposed recently by Prof. Huxley includes three principal subdivisions—viz.: *Prototheria*, *Metatheria*, and *Eutheria*. The terms *Prototheria* and *Eutheria* were employed by Prof. Gill a number of years previously to designate the two principal groups of this class, and appear to have been appropriated by Huxley without credit.

considered of sufficient value by some to give the *Didelphia* a rank equal to that of the *Prototheria*, and they consequently make three primary divisions of the class—*Ornithodelphia, Didelphia,* and *Monodelphia,* after De Blainville. If this were associated with any other characters of structural importance it would be quite sufficient, but since it is not, and in view of its unreliability and inconstancy in the lower Vertebrata, I am not disposed to regard it as equal in value to the strong structural characters by which the *Prototheria* are defined.

The subdivision of the Monodelphia is not an easy matter, if indeed any important divisions further than the separation of the mutilate series can be made. It is convenient, however, to adopt the classification of Lamarck, and divide them into three series, as follows: the mutilate series, the ungulate series, and the unguiculate series. The first of these includes the *Cetacea*, or whales, and the *Sirenia*, or sea-cows. The only character by means of which they are associated is the absence of hind limbs and the loss of the articular processes of the bones of the manus. The *Cetacea* form a perfectly natural and homogeneous group, and are entitled to a wide separation from all other Mammalia. We at present know very little concerning their development or ancestry, further than that their Eocene representative, *Zeuglodon*, resembled the ordinary monodelphous type more than does any other member of the order. They are undoubtedly a very old and distinct group, and it would not be at all surprising if they are ultimately found to have descended directly and independently from the *Prototheria*.

The *Sirenia*, or sea-cows, on the other hand, appear to be simply modified ungulates that have gradually assumed their present structure in accordance with their aquatic environment. The Miocene genus (*Halitherium*) of this order had distinct hind limbs, and in many ways resembled the primitive hoofed Mammalia. For this reason it is probably best to associate them with the ungulate rather than with the mutilate series, since they differ in almost every essential feature from the *Cetacea*, except in the loss of the posterior members.

The separation of the two remaining series, ungulate and unguiculate, depends entirely upon the distinctions to be drawn between "hoof" and "claw." If we contrast, for example, two such structures as the claw of the lion and the hoof of the horse, the distinctions are perfectly obvious, and we will experience no difficulty in recognizing the differences; but if we carefully trace the respective lines of ancestry of these two forms backward to the Eocene Period, we will find them converging to such an extent as to involve the hoof-and-claw question in almost hopeless confusion.

There are, however, two principal lines or stems which have terminated in the distinctly hoof-bearing mammals on the one hand and the claw- and nail-bearing on the other. The exact point at which these two lines converge has not as yet been satisfactorily determined, but it is undoubtedly true that they approached one another to a remarkable extent in the early Eocene. The ancestry of the entire ungulate series is indicated by the *Taxeopoda* of Cope, to whose persistent efforts and scholarly researches we are alone indebted for their discovery and description.

The primitive or central stem of this order is the *Condylarthra*, from which we pass by easy stages through the extinct genus *Meniscotherium* to the little hyrax or "coney," whose classification has long remained a puzzle to zoologists. From this group the extinct amblypods and elephantoid animals likewise came, while the *Perissodactyla* and the *Artiodactyla* are traceable directly to it.

The unguiculate series also has a generalized order, from which all the others radiate in different directions. This order has been called the *Bunotheria* by Cope, and exhibits a central axis in the sub-order *Insectivora*, the representatives of which are among the oldest of monodelph mammals. From the *Insectivora* we derive the *Creodonta*, a group of extinct insectivoro-carnivorous animals which terminates in the *Carnivora*. In another line come the lemuroids, monkeys, and man, while in still another are the *Cheiroptera* or bats, which are simply insectivores modified for flight.

One other order, the *Edentata*, or sloths, armadillos, ant-eaters, etc., remains to be accounted for. Some authors believe them to be affiliated with the unguiculate series, and to have sprung from the central insectivorous group. Palæontology has so far given us very few if any hints concerning the origin of this order, and it is probable that it will not be until the Eocene and Miocene Tertiaries of South America are more fully explored that any important information will be had upon this subject. At present I consider the evidence too meagre to hazard an opinion.

DIVISIONS OF THE MAMMALIAN DENTITION.—Many years ago Prof. Owen called attention to the fact that in many of the *Eutheria* there are two sets of teeth developed during the life of the individual—a deciduous or milk set and a permanent set—while in others but a single set appears. The former of these two conditions he designated by the term *diphyodont*, and to the latter he gave the name *monophyodont* dentition. It likewise so happens that generally, in those that have two sets (diphyodonts), the teeth in the various parts of the mouth are different in form and complexity, while in those that have but a single set (monophyodonts) the teeth are alike throughout. The diphyodont dentition is therefore, as a general rule, *heterodont*, that is, there are many kinds of teeth, and the monophyodont dentition is *homodont*, or all the teeth are alike. It was therefore originally supposed by Owen that diphyodont and heterodont and monophyodont and homodont were correlative and interchangeable terms, but it has since been discovered that there are many exceptions to this rule.

It must be borne in mind that the terms diphyodont and monophyodont are simply conveniences by which we are enabled to express briefly the conditions of replacement, and are not in any way to be looked upon as definitive of a natural group. The degree to which the second dentition is developed in the various sections of the Mammalia is subject to extreme variation, and it is not always an easy matter, if not frequently an utter impossibility, to determine whether certain teeth belong to the deciduous or permanent set, or in the monophyodonts to say whether it is the permanent or deciduous set which has been lost. There are, however, as will appear later, several important series in which the replacement and position are sufficiently constant to enable us to divide the teeth into several categories, the convenience of which, to say the least, if not the real importance, is undeniable. The question of the nature and relationship of the milk dentition to the permanent one will be discussed after the teeth of the several groups have been considered.

THE TEETH AND THEIR ACCESSORY ORGANS IN THE DOG.—I have thought best to next present a detailed description of the adult structure of an average diphyodont dentition, together with the accessory organs, in order that the student may become familiar with the technicalities before proceeding to consider the teeth of the various sections of the Mammalia. The dog has been selected as an example of this kind, partially on account of the generalized condition of the teeth, but principally on account of the readiness with which the student will be enabled to provide himself with the necessary material.

The teeth of the dog (Figs. 195 and 196) are forty-two in number, of which twenty belong to the upper and twenty-two to the lower jaw. The most anterior teeth of the upper series are relatively small, and are implanted in the free edge of the premaxillary bones in such a manner as to describe the arc of a circle. These are known as the *incisors* (*ic*, Figs. 195, 196). Behind these, after a slight interval, are

FIG. 195.

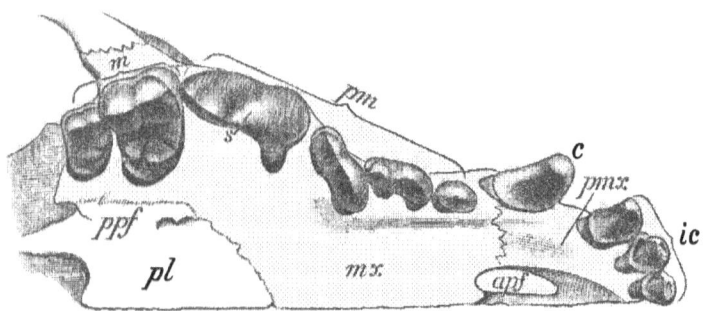

Vertical View of the Upper Jaw of a Dog (*Canis familiaris*): *ic*, incisors; *c*, canine; *pm*, premolars; *m*, molars; *s*, sectorial; *pmx*, premaxillary bone; *mx*, maxillary bone; *pl*, palatine; *apf*, anterior palatine foramen; *ppf*, posterior palatine foramen. The position of the third premolar is slightly abnormal.

a pair of strong, laterally compressed curved teeth, the *canines*, which are lodged deeply in the substance of the maxillary bone, immediately behind the maxillo-premaxillary suture. Behind the canines, again, are six teeth on each side, which progressively increase in size and complexity as we proceed backward until the penultimate tooth is reached, the last one being somewhat smaller. These are termed *molars* and *premolars*. The tooth-line of each moiety of the upper jaw presents three curves, the most anterior of which is formed by the three incisors and canine, with a strong convexity outward; the line of the next four describes a gentle curve whose convex surface is inward; while that of the last two curves boldly inward toward the median line.

The number of teeth in the lower jaw is one in excess of that of the upper, which is caused by the addition of a small single-rooted tooth at the posterior end of the series. They describe the same curves, so as to oppose those of the upper series. The incisors of the upper jaw, as has already been stated, are lodged in the pre- or intermaxillary bones, which limit the anterior part of the oral cavity above. The definition, therefore, of an incisor tooth of this series is *one which has a pre- or intermaxillary implantation irrespective of its size or form.* The incisor teeth of the lower jaw are *the corresponding ones which are brought into opposition with those of the upper jaw when the mouth is closed.* The teeth thus defined are three in number upon each side above and below in this animal, and are implanted by single slightly recurved fangs in distinct sockets or alveoli. In the upper series the median pair is the smallest, the outer ones gradually increasing in size. The base of the crowns of the four middle teeth is somewhat trihedral in form, with the apex flattened from before backward and produced into three cusps, of which the central one is the largest. The entire apex of the crown is slightly recurved. Upon its inner aspect the crown presents a basal ledge or cingulum, which sends out a low ridge to each of the lateral cusps. The lateral incisors are the largest and are somewhat caniniform. Like the median ones, their crowns have a strong basal cingulum posteriorly, but the lateral cusps are absent; the apex terminates in a strong hooked point.

The incisors of the lower jaw are similar to those of the upper, with the exception of the median pair, which is much the smallest and occupies a more anterior position than the others. The internal lateral cusps of these teeth are very faintly indicated, if indeed they can be at all made out, while the external lateral cusp is present and situated high up in the two median pairs. In the lateral ones it has a position nearer the base of the crown, and is separated from the median cusp by a deep fissure.

Between the lateral incisors and canines of the upper series there is a space or *diastema* about equal to the width of the lateral incisor. This space serves to receive the lower canine when the mouth is closed. At the back part of it upon the outside may be seen the suture by which

Fig. 196.

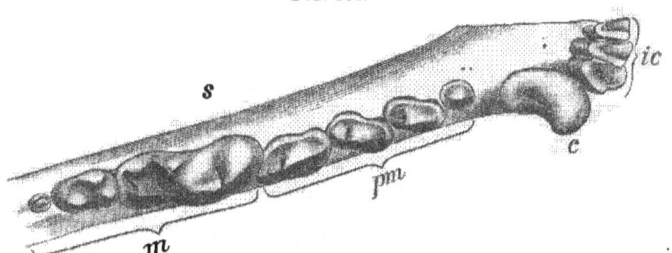

Vertical View of the Lower Jaw of a Dog (*C. familiaris*): *ic*, incisors; *c*, canine; *pm*, premolars; *m*, molars; *s*, sectorial.

the premaxillary bone joins the maxillary in the dentigerous border of the jaw. Just behind this suture the superior canine is lodged.

The definition, then, of a superior canine tooth is *one which is situated in the maxillary bone immediately behind the maxillo-premaxillary suture, provided it be not too far back, whatever may be its form, size, or func-*

tion, *while the canine of the lower jaw is the tooth which closes just in front of it.*

The canines of the dog are large, recurved, pointed teeth, projecting far above the level of the others, with slightly trenchant anterior and posterior edges. They are almost equal in size and very similar in shape. A very useful means by which they can be distinguished from each other, if at all worn and isolated from the rest of the teeth, is to note the point at which the worn surface exhibits itself. It must be remembered that the lower canine closes in front of the upper, in consequence of which the *posterior* face of the lower impinges against and abrades the *anterior* face of the upper; the anterior face of the lower canine also comes in contact with the lateral incisor, and an abrasion takes place at this point; but the posterior face of the upper canine is seldom worn except by long-continued use, so that ordinarily these points of wear serve as a useful guide in distinguishing between them. There is a slight difference in form, which can be ascertained only by close and careful comparison.

Behind the canines are four teeth which have been designated *premolars*. The reason for this distinction is founded upon the circumstance that these are the teeth situated behind the canine which vertically succeed the corresponding ones of the deciduous or milk set. The definition, therefore, of a premolar tooth is *one which, being situated behind the canine, displaces in a vertical direction a deciduous or milk tooth; all others behind these are true molars.* This is the definition which was originally proposed by Owen, to whom we are greatly indebted for this nomenclature: it would appear to be entirely satisfactory and sufficient, were it not for the fact that the first tooth counting from before backward, which is generally enumerated in the premolar series, *does not have any deciduous predecessor.* If we adhere strictly to this definition, it cannot be justly considered a premolar, but common usage has so long given it a place in this category that it appears advisable to still call it such. It should be remembered, however, that this is by no means an isolated case, but that other animals exhibit similar peculiarities.

The first premolar, so called, of the superior series is the smallest of the four, and is implanted rather obliquely in the maxillary bone; its single fang is slightly compressed laterally, and joins the crown at a moderately well-defined neck. The crown has an elongated oval form, terminated by a prominent obtuse cusp and surrounded by a well-marked ledge or cingulum, which is most conspicuous upon its inner face. From the summit of the main cusp two well-defined ridges descend to the cingulum, one on the posterior and the other upon the anterior border, giving to the tooth a slightly trenchant appearance. The hindmost of these two ridges divides the posterior half of the crown into two equal parts, and terminates with a very slight enlargement in the cingulum, while the anterior one has a more internal direction, and terminates in a distinct tubercle which occupies a position at the base of the antero-internal portion of the crown. The two ridges and the cingulum below enclose a shallow triangular depression internally, the outer face being convex.

The second and third premolars are considerably larger than the first, and are implanted by two roots, of which the posterior is the larger. These two teeth resemble one another very closely, the only appreciable difference being their slight disparity in size. Their crowns, like that of the first, are of greater longitudinal than transverse extent, and are produced into a prominent cusp situated a little anterior to the centre. The posterior ridge is interrupted shortly before it joins the cingulum by a deep transverse notch which gives rise to a distinct cusp, the *posterior basal tubercle*, situated over the hinder root. A faint indication of a second cusp is seen just behind this as an elevation of the cingulum. The antero-internal tubercle is present, and occupies relatively the same position as it does in the first premolar. The cingulum is more prominent on the inner than on the outer side of the crown, and with the two ridges encloses a triangular space.

The fourth premolar is by far the largest and strongest tooth of the premolar series. It is commonly known as the "flesh tooth," or *superior sectorial*, for reasons presently to be given. It is implanted by three roots, two external and one internal. The crown is composed of two principal lobes supported by the two external roots, and a small antero-internal one supported by the internal fang. The two principal lobes have an antero-posterior position, and are separated from each other by a deep, narrow fissure. Of these, the anterior is the larger and higher of the two; when viewed externally it resembles a cone with the anterior contour produced. Internally it is flattened somewhat, so as to correspond with the flattened inner surface of the posterior lobe. Posteriorly it is produced into a strong blade-like ridge, which is terminated by the vertical fissure, while its anterior surface is marked by a moderate vertical ridge. The posterior lobe is essentially chisel-shaped in form, with the bevelled edge external; its apex forms a blade-like crest which extends the entire length of the lobe. The internal lobe is small, and occupies a position at the antero-internal angle of the crown, being connected with a faintly-marked cingulum which surrounds the base of the crown. When we attempt to homologize the component lobes of this tooth with those of the premolars in advance of it, it is not difficult to see that the anterior lobe is the principal cone, that the posterior one is merely an exaggerated posterior basal tubercle, while the internal lobe is strictly homologous with the structure which has a similar position in the others. The three anterior premolars are not in as close contact as the teeth in the back part of the jaw, but are separated from each other by slight intervals, which are most conspicuous between the first and second.

The premolars of the lower jaw are similar in form to those of the upper, with the important exception of the fourth or last, wherein there is to be found a wide difference both in size and structure. The first of the lower series is smaller than the corresponding tooth above, and has a simpler, more conical crown. It is separated by a considerable diastema from the canine in front of it, but is almost in contact with the second behind. The second and third resemble those which are in a like position in the upper jaw, while the fourth is also similar to the corresponding tooth above, with the exception of a slightly increased size and the possession of a well-defined second posterior basal lobe. It slightly overlaps the great first true molar behind it.

The true molars of the superior series are two in number upon each side, and are placed directly behind the premolars. The definition of a true molar tooth is *one which, being situated behind the premolars, does not displace a deciduous or milk predecessor.* The two molars above are three-rooted, with broad tuberculated crowns imperfectly quadrangular in outline. The first, which is more than twice the size of the second, has two strong obtuse conical tubercles on the external portion of the crown, situated directly over the anterior and posterior external roots; they are subequal and separated from each other by a transverse notch. Internal to these there is a broad ledge, well rounded off internally, bearing three cusps. The one most internal is lunate in form, and is closely connected with the cingulum, which surrounds the base of the tooth. The cusp situated near the antero-internal angle is the largest, and has a subtrihedral form. A distinct ridge passes outward and forward from it to join the cingulum. Posterior to this last-mentioned cusp, and separated from it by a wide open notch, is the third tubercle, less distinctly marked than either of the others. An analysis of the various cusps of which the crown is composed leaves little room to doubt that the two external cusps are strictly homologous with the two external ones of the sectorial in advance of it—that the internal ledge which bears the three tubercles represents the greatly enlarged internal lobe of the sectorial, which has been removed to a more posterior position, and has acquired an important addition from the cingulum. That part of it which is exactly homologous with the internal lobe is the principal cusp at the antero-internal angle, which in some carnivorous animals is continued outward and backward as a prominent ridge, and does not develop the third tubercle. If the lunate cingular cusp be subtracted, the crown will be seen to resemble that of the sectorial in its general features.

The second true molar is like the first, except that the internal ledge exhibits, instead of three tubercles, two crescentiform ridges.

The first true molar of the lower jaw is the largest tooth in the entire dentition of the dog, and is the *sectorial* of the inferior series. It is implanted by two powerful roots at a point about midway between the anterior extremity and the condyle of the lower jaw, and occupies a position near the canthus or angle of the mouth. Its crown may be described as composed of two anterior blade-like cusps, a small internal tubercle, and a low tubercular heel. Of the two anterior cusps, the posterior is the larger, and rises gradually above the level of the one anterior to it; both are convex internally, but somewhat flattened externally to correspond with the internal flattened surface of the two blades of the superior sectorial. They are separated from each other by a deep, narrow fissure. The heel is low, and occupied by two cusps disposed transversely, of which the outer one is the larger. A faint ridge connects them, enclosing a shallow basin in front; on this account the heel is said to be basin-shaped. The internal tubercle is small, and is placed at the inner posterior part of the median lobe. In many carnivores it completely disappears, as does also the heel, as we shall presently see. The fourth superior premolar and the first inferior true molar are called *sectorial*, on account of their scissor-blade structure and the manner in which they oppose each other. If the macerated skull of a dog be carefully examined, it will be seen that the incisors of each series oppose each other almost exactly, while the lower canines close in front of the upper. As a consequence of this, the first premolar below closes in advance of the first premolar above; the second below in the interval between the first and second above, etc., but always upon the inside, on account of the unequal width of the two jaws. Now, the inferior sectorial bites against the superior in such a manner that its blades exactly oppose those of the tooth above after the manner of a pair of shears, so that when the mouth is closed the inferior sectorial is completely hidden from view; the heel opposes the first true molar above. Those who have ever studied the habits of dogs or wolves must have noticed that when they wish to divide a tough animal membrane or ligament they pass it back to the canthus of the mouth on one side and make several short quick strokes of the jaw; this is the shearing movement of the sectorials.

The remaining two true molars are much smaller, the last being one of the smallest of all the teeth, and is implanted by a single root. It is said to be permanently absent in some races of the domestic dog, especially the "pugs" and "Japanese sleeve dogs." The second molar is two-rooted, with a tuberculate crown of a more or less quadrate form. Two transverse cusps occupy the anterior part, while a third is placed at the postero-external angle of the crown on the edge of a broad flat heel. A basal cingulum is also present. The last tooth has an obtusely

Fig. 197.

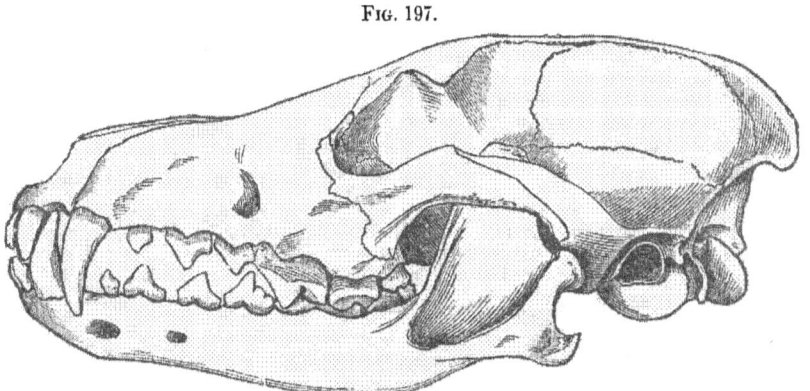

Side View of the Skull of a Dog (*C. familiaris*).

conical crown. The homologies of the cusps of the inferior true molars are not evident in the dog, but when we come to examine allied forms it will be found that the two blades of the sectorial represent the primitive cone, and the anterior basal lobe of the ordinary premolar greatly enlarged and specialized, while the heel represents the two posterior basal tubercles arranged transversely; the internal tubercle is an extra outgrowth from the cingulum.

In the case of many extinct animals the succession, and consequently the discrimination, of the molars and premolars would be attended with considerable difficulty were it not for the fact that in a majority of the Mammalia the first true molar is the first of the permanent set of the molar and premolar series which comes into place. By the time the last or fourth premolar is cut, which is usually one of the last, the first true molar immediately behind it is considerably worn down by use, so that this disparity of wear will of itself frequently serve to locate the exact limits of each series. It is a rule which is often employed by palæontologists to determine the dental formula of an animal the succession of whose teeth is unknown. When the anatomist wishes to indicate briefly the number of the various teeth of any particular animal, he employs what is called a dental formula. By this method the permanent dentition of the dog would be expressed as follows: I. $\frac{3}{3}$, C. $\frac{1}{1}$, Pm. $\frac{4}{4}$, M. $\frac{2}{3}$; which means that there are three incisors upon each side above and below, that there is one canine upon each side above and below, that there are four premolars, and that there are two true molars above and three below. This manner of abbreviation is convenient and easily understood, and saves both time and space in descriptions.

The division of the teeth into incisors, canines, premolars, and molars, although open to some objection, is nevertheless useful, since it serves to locate, in the case of addition or subtraction of a tooth to or from the normal diphyodont number, the exact position in which the change has taken place. In the marmoset monkeys of South America, for example, the total number of teeth is thirty-two, the same as in man. An inspection of their formula, however, which is I. $\frac{2}{2}$, C. $\frac{1}{1}$, Pm. $\frac{3}{3}$, M. $\frac{2}{2}$ = 32, will show that there is an important difference between the number of molars and premolars, the formula in man being I. $\frac{2}{2}$, C. $\frac{1}{1}$, Pm. $\frac{2}{2}$, M. $\frac{3}{3}$. In the former it is a molar which is lost; in the latter it is a premolar. Another example of this kind is seen in the upper and lower teeth of the otter, in which they are equal in total number, but unequal as far as the respective kinds are concerned. The dental formula in this animal is I. $\frac{3}{3}$, C. $\frac{1}{1}$, Pm. $\frac{4}{4}$, M. $\frac{1}{2}$ = 36.

The Accessory Organs.—This subject properly embraces a consideration of the bones by which the teeth are supported, the muscles concerned in their movement, the blood-vessels by which they are supplied with nutriment, and the nerves distributed to them. The bones in which the upper teeth are implanted are the *maxillæ* and *premaxillæ*, which are usually enumerated as bones of the face. The maxillary bone (Fig. 195, *mx*) is by far the largest one belonging to this category, and forms the greater part of each moiety of the upper jaw. It likewise contributes to the formation of the cheek, orbit, and palate, and also takes the principal share in forming the boundary of the nasal chamber. The maxillary bones do not meet in the median line above, on account of the interposition of the nasals and premaxillaries, but below they send inward two thin horizontal plates which meet in the middle of the roof of the mouth.

For descriptive purposes it is convenient to divide each bone into three external surfaces, the facial, palatine, and orbital. The facial surface, which is the largest of the three, is directed outward, and is irregularly triangular in form, with the apex directed forward. The superior border of this surface is considerably curved, and joins the premaxillary in front and the nasal behind. The posterior border is irregular, and is in contact from within outward with the frontal, lachrymal, and malar bones respectively, being excluded by these bones from the rim of the orbit. The inferior border is known as the dental or alveolar border, on account of its affording support to the canine, premolar, and molar teeth of the upper jaw. It is in contact with the premaxillary in front, and terminates behind in a free extremity beneath the orbital fossa.

The surface thus bounded is uneven, being interrupted by elevations and depressions. At the anterior angle the superior canine is implanted, and the course of its powerful curved root is indicated by a well-marked ridge or swelling of the external surface. Behind and above the posterior termination of this is a broad, shallow depression, while behind and below is another depression, the *canine fossa*, ending posteriorly in a large foramen, the *infraorbital foramen*, situated above the interval between the third and fourth premolars. Behind the infraorbital foramen a strong process is thrown up to meet the malar; this is known as the *malar process* of the maxillary. The posterior superior angle is produced into a considerable rounded process, which passes as far backward as the centre of the orbit to articulate with the frontal. This is the *nasal process* of the maxillary, and is the homologue of a corresponding process in the human skull bearing this name.

The posterior or orbital surface is relatively small, convex from before backward, and concave from side to side. It is somewhat triangular in shape, and forms the greater part of the floor of the orbit, being directed upward and backward. It is bounded above and externally by the malar, directly above by the lachrymal, and internally by the palatine bones respectively, terminating in a free rounded border behind. The internal portion of this last-mentioned border is separated from the palatine by a notch, and forms a conspicuous eminence known as the *maxillary tuberosity*. At the anterior extremity of this surface is seen the posterior opening of the infraorbital canal, which traverses the maxillary bone and serves for the transmission of the second division of the trigeminal or fifth nerve, as well as a part of the external carotid artery, which terminates in this situation as the infraorbital. This surface is perforated by small foramina for the entrance of the superior dental nerves and arteries.

The inferior or palatal surface forms a considerable part of the bony roof of the mouth as well as the floor of the nasal chamber. It is limited in front by the premaxillary bone, externally by the free alveolar border, posteriorly in part by the palatine and in part by a free edge, and internally by the suture with which it joins its fellow of the opposite side. It is slightly concave from side to side, the alveolar border being considerably elevated. Posteriorly it sends backward a narrow strip which terminates in a free edge behind; anterior to this, at a point opposite to the anterior part of the sectorial, it widens rapidly. From this point to its anterior termination it gradually narrows again. Just internal to the sectorial is seen a deep depression, the *sectorial fossa*, which serves to accommodate the blades of the inferior sectorial when the mouth is closed. Internal to this, again, are usually two, sometimes three, foramina, the *posterior palatine foramina*, which transmit the posterior palatine vessels and nerve. From the largest, most anterior of these a shallow groove is continued forward in which the palatine artery is lodged. This is the *palatine groove*.

The line of junction of the two palatal plates of the maxillaries is marked by a longitudinal ridge, the *sutural ridge*, which gives support to the vomer above. The maxilla articulates with the premaxilla and nasal in front, with the frontal above, and with the lachrymal, malar, and palatine behind.

The premaxillæ are small bones placed in front of the maxillæ, the two together forming the anterior termination of the upper jaw. Each consists of a thickened anterior portion meeting in the median line, together with an ascending or vertical process and a horizontal process. The thickened body forms the lower boundary of the anterior nares, and by its free alveolar border lodges the incisor teeth. The ascending or vertical process is a long, sharp spicule of bone which springs from the outer side of the body and furnishes the external wall for the narial opening. It is directed upward and backward, and insinuates itself between the nasal above and the maxilla below. This is known as the *nasal process* of the premaxilla. Upon either side of the median line the horizontal or *palatine processes* pass backward to the maxillæ, forming the anterior portion of the bony palate. These processes are in contact with each other in the middle line, but each is separated from its body by a wide hiatus, which is converted into a foramen by the interposition of the palatal plate of the maxillary behind. These large oval foramina are conspicuous features in the macerated skull, and are known as the *incisive* or *anterior palatine foramina*. They transmit the anterior palatine vessels and nerve.

The next and last bony structure to be noticed in connection with the teeth is the *mandible* or lower jaw. This part of the skeleton in human anatomy is known as the *inferior maxilla*, and consists, in the adult state at least, of a single bone (the two halves co-ossified), as is also the case in the monkeys and several other mammals. In the majority of them, however, it is made up of two more or less persistent pieces, which may unite in extreme old age to form a single bone. The mandible of the dog consists of two symmetrical elongated halves, *the rami*, diverging behind and coming in contact in front in the median line by two roughened surfaces, the *symphysis*. They are bound together by the interposition of fibro-cartilage at this point, and are movably articulated to the skull behind by two transversely elongated processes, the *mandibular condyles*, placed near the middle of the posterior border. Each ramus is laterally flattened, with the inferior border considerably curved in an antero-posterior direction. In front this border slopes gradually upward to meet the alveolar or dentary border, while behind it is terminated by a prominent, slightly-inflected process, the *angle*. The dentary border is nearly straight, and is prevented from reaching the posterior border by the intervention of a broad flat recurved plate of bone, the *coronoid pro-*

cess, which rises high above the level of the surrounding parts. The posterior border is interrupted by two notches, between which is situated the condyle. Immediately in front of the condyle is a wide and deep depression, the *masseteric fossa*, for the insertion of the powerful masseter muscle. In front of and below the condyle, on the inner side, is a conspicuous opening, the *inferior dental canal*, which gives passage to the inferior dental artery and nerve. On the external surface, behind and below the root of the canine, is another opening, the *mental foramen*, through which a part of the nerve makes its exit to be distributed to the lower lip.

The Muscles.—The principal muscles concerned in the movement of the lower jaw are the temporal, masseter, external, and internal pterygoids, the digastric, genio-hyoid, and mylo-hyoid.

The *Temporal* is a broad, thick, fleshy muscle which covers the side wall of the brain-case from the post-orbital process in front to the lambdoidal or occipital crest behind, reaching as high up as the sagittal crest above, and completely filling up the temporal fossa, to which it gives its name. Its fibres converge fan-wise to be inserted into the summit of the coronoid process of the ramus of the mandible. Its principal action is to elevate the lower jaw. By its leverage and great strength the animal is enabled to take a firm grip upon its prey.

The *Masseter* is a short, thick muscle arising from the under and a part of the outer surface of the malar bone, as well as the posterior part of the maxillary, and, passing downward and backward, is inserted into the masseteric fossa of the ramus. Its action is similar to that of the preceding muscle.

The *Internal Pterygoid* muscle consists of a strong bundle of muscular fibres which takes its origin from the pterygoid fossa in the base of the skull, and passes downward and outward to its insertion in the lower part of the angular process. By its contraction the lower jaw is drawn upward and inward, but owing to the manner in which the teeth interlock no extensive lateral movement is possible. The most reasonable view of the action of this muscle, as well as the succeeding one, is, that by the contraction of those of one side the sectorial apparatus of the side opposite is enabled to perform a more perfect shearing movement, just as the blades of a pair of scissors must be pressed closely together in order to make them cut. From the direction of its fibres it likewise assists in elevating the jaw.

The *External Pterygoid* arises from the pterygoid plate of the sphenoid bone, and is inserted into the base of the condyle, and as far forward as the inferior dental canal. Its action has already been alluded to.

The *Digastric* is a large muscle which arises from the skull behind the auditory bulla in a strong bony prominence, the paramastoid process, and passes forward to its attachment on the inferior margin of the ramus in front of the angular process. Its action is to depress the jaw and open the mouth.

The *Genio-* and *Mylo-hyoid* muscles are broad muscular sheets which lie between the rami forming the floor of the mouth in the recent state, being attached to the hyoid bones and the "fork" of the jaw. They assist in depressing the mandible, and consequently in opening the mouth.

Vessels and Nerves.—The blood-vessels by which the teeth and the muscles described above are supplied are derived from the external carotid artery, which passes forward along the side of the neck, giving off branches to the various structures in this situation. This artery does not terminate, as in man, in the temporal and internal maxillary arteries —at least it is so generally considered by anatomists. Both the right and left common carotids spring from the innominate, as in the *Carnivora* generally. After giving off the thyro-laryngeal branch, remarkable for its large size, to the thyroid gland and larynx, it passes forward in front of the transverse process of the atlas vertebra, where it gives off the occipital artery, which goes to the back of the head and the deep muscles of the neck. Upon the base of the skull in the vicinity of the carotid canal it bifurcates into two principal branches, the external and internal carotids, the latter entering the skull through this canal to be distributed to the brain, the latter continuing forward through the alisphenoid canal, giving off in its course the laryngeal, lingual, facial, posterior auricular, and superficial temporal branches. Near the condyle of the lower jaw it describes a remarkable sigmoid curvature between this structure and the internal pterygoid muscle, thence passing forward to the infraorbital canal, where it receives the name of the infraorbital artery. Between the condyle and the infraorbital canal the following principal branches are emitted by this arterial trunk: the inferior dental artery, which enters the inferior dental canal and supplies the teeth of the lower jaw; the deep posterior temporal, which furnishes a masseteric branch passing through the sigmoid notch, or that between the condyle and coronoid process of the ramus, to enter the masseter muscle; several pterygoid arteries, which go to the pterygoid muscles; the ophthalmic artery, distributed to the eye; the deep anterior temporal; the palatine, buccal, and alveolar arteries; lastly, the superior dental artery, which supplies the teeth of the upper jaw.

The nerves supplying the teeth and accessory organs are derived principally from the trigeminal or fifth pair of cranial nerves. This is essentially a mixed nerve in function, arising by two roots, a large sensory and a small motor root. At a short distance from its origin the sensory root swells out into ganglionic enlargement, the Gasserian ganglion, after which it divides into three branches—the *ophthalmic* or *first division*, the *superior maxillary* or *second division*, and the *inferior maxillary* or *third division*.

The first of these makes its exit from the cranial cavity through the sphenoidal fissure, and supplies by its subdivisions the eyeball, mucous membrane of the eyelids, the skin of the nose and forehead, dividing into frontal, lachrymal, and nasal branches.

The second of these branches, the superior maxillary, issues from the skull through the foramen rotundum, and supplies the side of the nose, upper teeth, and the upper part of the mouth and pharynx. It crosses from the foramen rotundum directly to the infraorbital canal, in the vicinity of which it gives off the anterior and posterior dental nerves which supply the teeth.

The third division, inferior maxillary, passes out of the skull through the foramen ovale, just outside of which it is joined by the motor root. It then divides into two branches, a small anterior one distributed to the muscles of mastication, and a large posterior branch, which supplies the ear, side of the head, lower lips, gums, teeth, salivary glands, and inside of the mouth. The posterior branch divides into the *auriculo-temporal*, which passes backward to the temporal region; the *inferior dental*, which supplies the teeth of the lower jaw; and the *gustatory*, or the nerve of taste, which goes to the mucous membrane of the tongue.

The lips, tongue, and salivary glands should also be mentioned in connection with the accessory organs, since they serve an important purpose in preventing small particles of the food from escaping during mastication, as well as supply the requisite moistening fluid whereby comminution is more readily accomplished and the food rendered more digestible.

TEETH OF THE EDENTATA, OR BRUTA.

Although this group is by no means the most primitive of the Mammalia, as will be seen by reference to the table of classification, yet the characters which we have assigned to the ideal primitive mammalian dentition are most nearly approached in certain members of this order. Whether the comparatively simple form and absence of enamel in the adult tooth, which is characteristic of all the animals of this order, pertain to a primitive state, or whether this condition has been reached by a process of retrogression or degradation, as many believe, we are not at present prepared to say, in the absence of any knowledge of their palæontological history beyond the latest Tertiary epoch. There is one character, however, in which one at least is more decidedly primitive than any other known Eutherian mammal, and that is *the succession of all the teeth but one (the last) by a second set*. I refer to the nine-banded armadillo (*Tatusia peba*). It is not certainly known whether this condition exists in any other of the edentates or not, with the exception of the sloths, which are truly monophyodont.

The term *Edentata* is inappropriate, inasmuch as one would be led to suppose from the name that they have no teeth. The original term, *Bruta*, was applied to this order by Linnæus, which he defined by the absence of incisor teeth. It was afterward changed to *Edentata* by Cuvier—a name which has been extensively adopted by subsequent authors. It was formerly supposed that no incisor teeth are ever present in this group, but the discovery of new forms proved this to be erroneous. The median incisors, however, are wanting in all cases so far known. The definition of the order now most commonly given is "absence of enamel on the teeth." This peculiarity appears at first sight striking and quite sufficient to separate them from all other monodelphous mammals, but C. S. Tomes has shown[1] that the tooth-germs of the nine-banded armadillo have distinct enamel organs, which are subsequently aborted as the tooth comes to maturity. This discovery is important, since it indicates pretty clearly that the loss of enamel is a

[1] *Philos. Trans.*, 1876.

TEETH OF THE VERTEBRATA.

mark of degeneracy, and leads indirectly to the conclusion that the armadillos at least are descended from ancestors with enamel-covered teeth, who in all probability were the possessors of a completely developed second set.

The only assignable cause for this degenerate condition of the dental organs is the peculiarity of their food-getting habits. Many of them feed upon insects, which they capture by means of a long whip-like tongue covered with the viscid secretion of the submaxillary glands, and swallow whole. This manner of feeding would occasion little demand for masticatory organs, which from disuse would gradually fall into a rudimentary condition and eventually disappear. Those in which the entomophagous habit is exhibited in its greatest perfection are edentulous, and have small mouths with extremely long tongues. All the Edentata in which this structure exists at all show a tendency toward such a habit—even the arboreal sloths, which are said to be exclusively vegetable feeders.

Flower has recently shown that the sloths are intimately connected with the ant-eaters and armadillos of South America through the extinct megatheroids, and that all the American forms have probably descended from a common ancestor, while the Old World forms are likewise closely related and descended in another line. It is probably true that the armadillos are most nearly related to the ancestral form, and that the sloths represent an offshoot which was derived from them after they had lost the enamel of the teeth in the manner indicated.

The teeth of the armadillos are, with one exception, relatively small cylindrical bodies implanted in the dentigerous borders of the lower jaw, maxillary, and sometimes premaxillary bones. They are entirely devoid of enamel, and grow continuously throughout the life of the animal, in consequence of which no roots are formed.

FIG. 198.

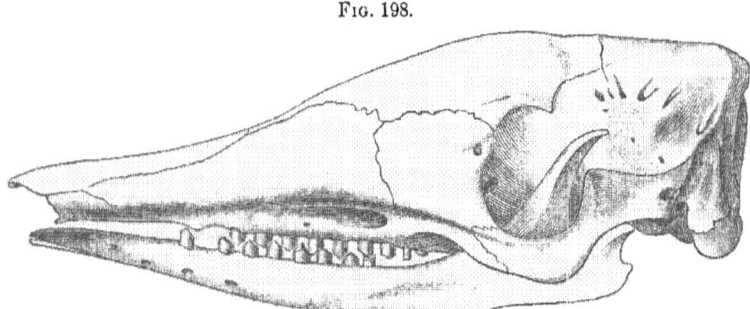

Side View of the Skull of a Seven-banded Armadillo (*Tatusia hybridus ?*)

In the seven-banded armadillo (*Tatusia hybridus*, Fig. 198[1]) there are seven teeth above and eight below upon each side. So little is known

[1] The specimen here figured is in the U. S. Army Medical Museum, and is labelled *Tatusia septemcinctus*. It exhibits the peculiarity of having eight teeth upon one side and seven upon the other in the upper jaw. There is, however, a considerable space between the first and second tooth of the right side, which would indicate that a tooth is missing. The number ascribed to this species by Owen is seven above and eight below upon each side. Its exact identification is therefore difficult.

about their succession that it is impossible to say whether there are molars and premolars represented or not. The teeth of the upper series are lodged in the maxillary bones, and begin at a considerable distance behind the maxillo-premaxillary suture. They progressively increase in size up to the fifth or sixth tooth, the last being quite small. They are not in contact with each other, but are separated by slight spaces about equal to the width of a tooth. The teeth of the lower jaw are similar to those of the upper jaw in size and shape, with the exception of the last, which is much larger than the corresponding tooth above. The teeth of the inferior series close in the intervals between those of the upper and conversely, causing the summits of the crowns to wear, as Prof. Owen puts it, "into two facets divided by a median transverse ridge." The form of the working surface of the tooth is therefore wedge-shaped. The first two teeth of the lower jaw shut in front of the first tooth above, and the last three teeth above behind the last one of the lower series, leaving them with little or no opposition. Each tooth continues its cylindriform shape to the bottom of the alveolus in which it is implanted, having its base excavated into a large pulp-cavity. It consists of dentine and cementum only.

In another species, the nine-banded armadillo, the number and form of the teeth are the same. The teeth of this animal, as has already been stated, have a successional set. According to the definition laid down for premolar and molar teeth in the diphyodont Mammalia generally, there would be one molar and six premolars in the dentition of this animal. The rooted appearance of the deciduous teeth, according to Tomes, is not due to the possession of true roots, but to the absorption set up by the approach of the successors.

The genus *Priodon* of this group has as many as one hundred teeth, the greatest number exhibited by any land mammal. They are relatively small and simple in form, and are confined to the maxillary and mandibular bones. They vary in number from twenty-four to twenty-six upon each side in the upper, and from twenty-two to twenty-four upon each side in the lower jaw. In the living genus *Dasypus* there is one tooth upon each side implanted in the premaxillary bone, which, according to the definition, becomes an incisor, while in still another extinct genus, *Chlamydotherium*, almost equalling in size the rhinoceros, there were two incisors above and three which opposed them below. In *Glyptodon* the teeth are more complex in pattern, being laterally compressed and divided by two vertical grooves upon each side, which are opposite to each other. The resulting structure from this arrangement is three transverse vertical plates connected in the centre by an isthmus. There were teeth in the premaxillaries in this genus.

The megatheroids afford another example of moderate complexity in the enamelless teeth of the *Bruta*. In the gigantic extinct *Megatherium* there are five molars above and four below upon each side. They are very deeply implanted in the substance of the jaw bones, and have remarkably elongated pulp-cavities, which communicate with the grinding surface by means of a narrow fissure. The pulp-cavity is immediately surrounded by soft, more or less vascular dentine—the vasodentine of Owen—which is covered by a thin layer of unvascular, much harder dentine. Upon the outside of this comes the cementum, which has a great thickness upon the anterior and posterior face of the tooth. Owing to the unequal powers of resistance which these substances offer, the teeth wear in such a manner as to present two transverse crests each, and are therefore spoken of as *lophodont*. They are confined to the maxillary and mandibular bones and grow from persistent pulps.

In another extinct allied genus, *Megalonyx*, the teeth are oval in section, and did not wear into transverse crests as in *Megatherium*, but have slightly concave grinding faces. The first tooth of the upper series also is considerably enlarged and caniniform in shape, as in one of the living sloths. Another nearly-related form is *Mylodon*, likewise extinct, which exceeded the rhinoceros in size. The first tooth above, instead of being enlarged and caniniform, is smaller than the succeeding ones, and otherwise like them in pattern.

The dental formula of the three-toed sloth (*Bradypus tridactylus*) is I. $\frac{0}{0}$, C. $\frac{0}{0}$, M. $\frac{5-5}{4-4} = 18$. It has been observed, however, that there is in some young examples of this species a small extra tooth in the lower jaw just in front of the first permanent one, which is shed before the animal attains to the adult state. The teeth are relatively small, of a columnar form, and implanted to a moderate depth in the substance of the jaws by a deeply-excavated base for the accommodation of the persistent pulp. The grinding surface presents a central depression in the vaso-dentine, surrounded by a raised rim on its outer margin composed of the harder dentine, which usually wears unequally into one or two prominent points. The teeth of the upper and lower jaws do not oppose each other exactly, but alternate when the mouth is closed.

In the two-toed sloth (*Cholœpus didactylus*) the dental formula is the same. The first tooth in each series, which in the edentates generally is the smallest, is here greatly increased in size, of a subtrihedral form, and of a caniniform pattern. They are separated by a considerable diastema from the rest of the teeth, and are implanted above in the maxillary bones a short distance behind the maxillo-premaxillary suture. It will be seen, therefore, that as far as the definition of a canine tooth is concerned, all the conditions are fulfilled; but the tooth in the lower series, which has undergone a similar modification, violates the definition of a canine, inasmuch as it closes behind the upper caniniform tooth instead of in front of it. It is therefore a matter of uncertainty whether these teeth are strictly homologous with the canines of the diphyodont Mammalia or not. From the manner in which they oppose each other the posterior surface of the upper and the anterior surface of the lower are extensively abraded, and their summits worn into sharp points, which would render them efficient weapons of offence or defence should the animal choose to use them as such.

The next tooth of the upper series is relatively small, and is implanted rather obliquely, with the summit inclined backward and inward. The two following teeth are larger, with a central depression upon the grinding face, and having the external and internal portions of the rim produced into sharp points. The last tooth is about equal to the second in size, which it also resembles in form.

The teeth of the lower series resemble those of the upper, with the exception that the three posterior ones are more robust and gradually decrease in size from the second to the last. Viewing the teeth and accessory bony structures of this animal as a whole, the premaxillæ are remarkable for their small size, little extension anteriorly beyond the maxillæ, and the complete absence of the ascending or nasal process, as well as their edentulous condition. The palatal plates of the maxillæ are widest in front and gradually narrow posteriorly, causing the dental series of opposite sides to converge behind. In the lower jaw the two halves are completely co-ossified, as in monkeys and man; the anterior part of the symphysis is produced into a peculiar spout-like termination, at the base of which the jaw widens rapidly; and the rami are little divaricated posteriorly. The posterior teeth are implanted in a strong inwardly projecting ledge, in consequence of which the dentigerous border gradually approaches the median line as it proceeds backward. The mental foramen is placed below the interval between the third and fourth teeth near the middle of the ramus.

With respect to the teeth themselves, Owen gives the following common and constant characters of both recent and extinct sloth-like animals, which would include the megatheroids: "Teeth implanted in the maxillary, never in the intermaxillary bones; few in number, not exceeding $\frac{5}{4}-\frac{5}{4}$; composed of a large central axis of vascular dentine, with a thin investment of hard or unvascular dentine, and a thick outer coating of cement. To these, of course, may be added the dental characters common to the order *Bruta*—viz. uninterrupted growth of the teeth, and their concomitant implantation by a simple deeply-seated excavated base, not separated by a cervix from the exposed summit or crown."

Of the two Old World genera now living, but one has teeth. This is the aard-vark (*Orycteropus*), or, as it is sometimes called, the Cape ant-eater. Its dental formula is M. $\frac{7}{6}-\frac{7}{6} = 26$, of which the anterior ones of each series are not unfrequently wanting or concealed by the gum. The teeth of the superior set progressively increase in size from before backward up to the last tooth, which is smaller. They are oval in section, with the exception of the fourth and fifth, and have wedge-shaped triturating surfaces, like the armadillos. The fourth and fifth above and the last two below have two vertical grooves, one upon each side, which give to them an hour-glass shape upon section.

The teeth of this animal do not exhibit the customary excavated base of the Edentata generally, but are continued solid to the bottom of the sockets. Their minute structure is peculiar, and resembles that found in *Myliobates* among the elasmobranch fishes; the dentine is of the variety known as plici-dentine. This consists of a series of small vertical parallel tubuli which pass up from, and are virtually prolongations of the pulp-cavity. From these the dentinal tubuli radiate toward the periphery, just as they do from the single pulp-cavity of the human tooth already described. Owing to this peculiarity, Prof. Owen regarded the tooth of *Orycteropus* as an aggregate of many denticles, each with its proper pulp-cavity and dentinal tubes.

In Europe fossil remains of edentates are known from the Middle Miocene of Sansan in France and the upper Miocene deposits of Pikermi in Greece. Two genera, *Macrotherium* and *Ancylotherium*, have been described by Lartet and Gaudry, from feet and limb bones principally, nothing being known of the skull or teeth. In South America fossil remains of this order are very abundant in the Pampean or Pliocene deposits. Older deposits on the Parana River have furnished M. Ameghino with numerous forms which stand in ancestral relation to those of the Pampean beds, and which, it is interesting to observe, have more or less enamel on the teeth. In North America, Prof. Marsh has described a genus under the name of *Morotherium*, from the Loup Fork or Upper Miocene strata, from feet bones only. The teeth are not known.

TEETH OF THE CETACEA.

According to Dr. Theo. Gill's arrangement, the cetaceans are divisible into three groups or sub-orders, as follows: *Mysticete*, including the "whalebone whales;" *Denticete*, or the "toothed whales," and *Zeuglodontia*, a division which includes *Zeuglodon* and several other extinct genera. In the *Mysticete*, teeth are present in a fœtal state only, being absorbed before birth. This loss of the teeth is compensated for by the development of large corneous plates, the "baleen plates," which depend from the roof of the mouth. The more important of these are of a triangular form, and are arranged along each side of the palate in such a manner as to be transverse to the axis of the skull, the centre being occupied by smaller ones, also placed transversely. Altogether, they form by their extremities a vaulted surface into which the large tongue fits accurately, their edges being broken up into numerous stiff hairs which project into the mouth. The animal feeds by taking large quantities of water into the mouth and expelling it again through the nostrils ("spouting"); any small aquatic animals which may have been contained in it are entangled in the fringes of the baleen plates, and subsequently collected by the tongue and swallowed. It will thus be seen that the baleen acts as a sort of sieve or strainer, and is pre-eminently adapted to the capture of the small aquatic forms with which the sea in certain places literally swarms. Each baleen plate possesses a pulp situated in a cavity at its base, from which it is developed, and through which it is regenerated as fast as worn away. According to Tomes, each hair-like fibre has within its base a vascular filament or papilla; "in fact, each fibre is nothing more than an accumulation of epidermic cells concentrically arranged around a vascular papilla, the latter being enormously elongated. The baleen plate is composed mainly of these fibres, which constitute its frayed-out edges; and in addition to this there are layers of flat cells binding the whole together and constituting the outer or lamellar portion."

All the whalebone whales possess rudimentary teeth, or rather dentine and enamel organs, which undergo very little calcification before absorption sets in. In the fin-backs (*Balænoptera*) these dentine organs are simple in the front part of the mouth, bifid in the middle, and trifid in the back part of the jaw. They are placed in an open groove along the jaw, as in all other Mammalia at this stage of embryonic growth, and do not differ from them in any important particular.

In the second sub-order (*Denticete*) no baleen plates are developed; teeth are always present, and are more or less persistent. They are implanted by single roots, and are in some instances very numerous. No second dentition has ever been observed in any member of this group, and they are, so far as known, truly monophyodont. In the common porpoise of our coast (*Delphinus chymene*), which is an average example of this sub-order, the teeth are about ninety-four in number, and are lodged in the premaxillary, maxillary, and mandibular bones. They are implanted by single slightly enlarged fangs in ill-defined sockets incompletely partitioned off from each other, and in what at first sight seems to be a wide-open groove. Their crowns taper gradually to a sharp point, which is strongly incurved. The first two teeth in the upper jaw are small and implanted in the premaxillary bone, which furnishes a very small part of the dentigerous border of the upper jaw. Behind these the maxillary teeth rapidly increase in size up to the seventh or eighth tooth, after which they continue to the fifteenth or sixteenth almost equal in size, and then gradually diminish in size toward the posterior part of the jaw. The teeth of the inferior series are like those of the upper, except that the posterior ones are more robust. The jaws are remarkable for their great length and narrowness, and the arrangement of the bones of the face when compared with other mammals is also peculiar. The coronoid process of the lower jaw is obsolete.

In other members of the *Delphinidæ*—the dolphin, for example—the teeth are often as many two hundred, the greatest number exhibited by any mammal, or they may be reduced to a single functional tooth, as in the narwhal (*Monodon*). In this latter species four teeth are found in a fœtal state, but the two lateral ones are lost or absorbed before birth. In the male narwhal the left of the two anterior ones, which is placed in the premaxillary bone, grows from a persistent pulp and attains a length of ten or twelve feet. This formidable tusk is almost straight, and is marked by spiral ridges which wind forward from left to right. The corresponding tooth of the opposite side sometimes reaches a development equal to that of the left, but more frequently its growth is arrested and it remains buried beneath the gum, as do both in the female. Owing to this circumstance, it has been thought that it serves as a sexual weapon similar to the antlers of the deer, but until the habits of the animal are better known this explanation of its use must remain conjectural.

The great bottle-nosed whale (*Hyperoödon bidens*) is, to all outward appearances, edentulous, but careful examination reveals the presence of two, sometimes four, well-calcified conical teeth in the front part of the jaw, which remain more or less completely hidden by the gum. In addition to these, there are usually twelve or thirteen small rudimentary teeth imbedded in the gums of both jaws, which soon disappear.

In the sperm whale (*Physeter macrocephalus*) the exposed and functional teeth are confined to the lower jaw. These are about twenty-seven in number in each ramus, loosely implanted in a wide-open gutter, with the alveoli or sockets scarcely perceptible. They are at first sharply conical, but by attrition wear down into obtuse cones, biting into pits or cavities in the gums of the upper jaw. In this jaw there are a number of persistent rudimentary teeth concealed in the thick gums, one pair of which is exposed in the small pug-nosed sperm whale (*P. simus*).

In the dolphin of the Ganges (*Platynista gangetica*) the total number of teeth is one hundred and twenty-four, of which there are thirty upon each side above and thirty-two upon each side below. In the young animal their crowns are produced into sharp cones, but by attrition the posterior teeth are worn down to such an extent as to become molariform in shape. The last tooth in this species not unfrequently develops a double root or fang, and is the only example of the kind to be met with in living cetaceans.

The teeth of the dolphins of the Amazon are surrounded at the bases of their crowns by a well-marked ledge or cingulum, which throws up a strong internal cusp. On this account Dr. Gill elevates the genus *Inia* to the rank of a family.

The sub-order *Zeuglodontia* is extinct, and includes a number of fossil cetaceans, some of which are estimated to have attained a length of seventy feet. They differ from living representatives of the order in many important osteological characters, but not more prominently than in their dental organization. Besides being heterodont and having the posterior teeth implanted by two or three roots, *some, if not all, were diphyodont as well.* The exact extent of the replacement, however, is not fully known, but it is certainly true that two sets of teeth were developed. In *Zeuglodon cetoides* (Fig. 199), from the Claiborne

FIG. 199.

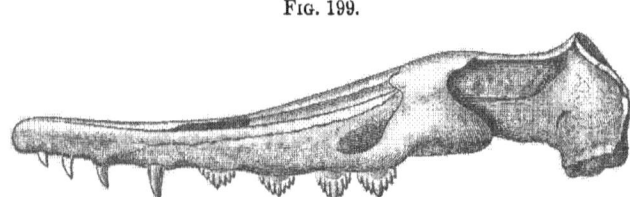

Side View of the Skull of *Zeuglodon cetoides* (after Gaudry).

Eocene deposits of Louisiana, Alabama, and Mississippi, the teeth are divisible by their form and position into incisors, canines, and molars. Three teeth with conical recurved crowns are implanted by single roots in each premaxillary bone, which in this animal contributes a considerable part of the tooth-bearing border of the upper jaw. Of these the anterior is the smallest and placed at some distance behind the extreme anterior border, the posterior ones gradually increasing in size. Behind these, near the maxillo-premaxillary suture, is a strong recurved single-fanged tooth of a caniniform pattern, and one which both by position and form becomes the homologue of the canine in the ordinary heterodont dentition. The rest of the alveolar border is occupied by four rather large more or less trenchant teeth, referable to the molar and premolar series. Each of these is implanted by two roots, and has a laterally compressed crown of a triangular form with the apex of the triangle at the summit. Each of the anterior and posterior edges are interrupted by three well-marked cusps, which give to the tooth a strongly-serrated appearance.

The heterodont and diphyodont character of the teeth of this cetacean serves to bring the anomalous, and in many respects degenerate, dental organs of this order into the closest relationship with the teeth of the ordinary diphyodont Mammalia, and, being the oldest member of the group so far known, goes far toward filling the wide gap between these aquatic and the terrestrial mammals.

TEETH OF THE UNGUICULATE SERIES.

In considering the dental organization of this vast assemblage of mammiferous animals it is necessary to have at the very outset a correct conception of the primitive or ancestral stock from which all of them have been derived, if such can be found to exist. This can be learned only by a careful study of the successional history of the various orders composing it. In searching, then, for this original stem we can by this method exclude many of the groups from this position by fixing the date of their appearance, and thereby establishing their exact limit in time. We know, for example, that the *Primates* could not be the ancestral group, for the obvious reason that they do not extend beyond Miocene time; nor the Carnivora, which appeared about the same time; nor the Rodentia, which date from the Middle Eocene; nor the *Cheiroptera*, which can be traced back no further than the Upper Eocene. We are therefore restricted in our choice to the insectivores, lemuroids, creodonts, or tillodonts, which alone of the entire series continue backward to the base of the Eocene Period. With reference to the creodonts, I do not believe that any important distinctions exist between them and the insectivores, while the line between this latter group and the lemuroids and tillodonts becomes extremely shadowy at this point.

Prof. Cope unites the insectivores, lemuroids, tillodonts, creodonts, and tæniodonts into one order, which he calls *Bunotheria*, and defines the several sub-orders as follows:[1]

I. Incisor teeth growing from persistent pulps:
   (a) Canines also growing from less persistent pulps, agreeing with external incisors in having molariform crowns . . . . *Tæniodonta*.
   (b) Canines rudimental or wanting; hallux not opposable . . . *Tillodonta*.
   (c) Canines none; hallux opposable . . . . *Daubentonioidea*.
II. Incisor teeth not growing from persistent pulps:
   (a) Superior true molars quadritubercular; hallux opposable . . *Proximiæ*.
   (b) Superior true molars quadritubercular; hallux not opposable. *Insectivora*.
   (c) Superior true molars tritubercular or bitubercular; hallux not opposable. *Creodonta*.

I believe, with this author, in classifying those forms in which the incisors grow from persistent pulps as a distinct group from those in which the incisors are normal, as far at least as their growth is concerned. In the first division there are three well-defined sub-orders. In the second division it is extremely questionable whether more than two sub-orders should be made. If we use the opposable and non-opposable condition of the hallux as a character, we will have two perfectly natural

[1] *Proceedings Academy Natural Sciences Philada.*, 1883.

series—the prosimian or lemuroid and the insectivorous; but if we go further, and establish another sub-order upon the tritubercular or quadritubercular character of the superior molar teeth, it will necessitate the wide separation of forms closely related to each other by every important feature of their anatomical structure—a course which I do not deem advisable nor in keeping with our present knowledge of the subject.

According to Prof. Cope's definition, the only character in which the Creodonta differ from the Insectivora is the tritubercular superior molars as distinguished from the quadritubercular; and in order to make the Creodonta homogeneous he is compelled to take out of the old group Insectivora the *Tupaiidæ*, *Centetidæ*, *Chrysiochloridæ*, and *Talpidæ*, and place them in the Creodonta. Aside from the inadvisability of such a course, these teeth in many of the above-named families are altogether intermediate between the tritubercular and quadritubercular pattern, the postero-internal tubercle being represented often by a rudimentary cingulum, which may be entirely absent or produced into a strong cusp. Then, again, the superior molar teeth of the prosimian division are indifferently tritubercular or quadritubercular; and if we adhere to this practice in the one, why not in the other? In consequence of these facts, I propose to unite the Creodonta with the Insectivora into a single division, for which the old name Insectivora may be retained.

Thus constituted, palæontological history, in my judgment, points strongly to the fact that this group stands in the important relationship of ancestors to a large part, if not the whole, of the unguiculate Mammalia. Working upon this hypothesis, it will be desirable to describe the more important types of dental structure to be met with in this sub-order, after which they can be followed out to their respective terminations in the various orders and sub-orders which make up the series.

INSECTIVORA.—The simplest form of dental structure in this sub-order is exhibited by the extinct genera *Mesonyx* and *Dissacus* of Cope, from the American Eocene strata. The teeth of *Mesonyx* (Figs. 200, 201) are forty-four in number, disposed as follows: I. $\frac{3}{3}$, C. $\frac{1}{1}$, Pm. $\frac{4}{4}$, M. $\frac{3}{3}$ = 42. The incisors are relatively small, with subconic crowns, which are closely approximated. The superior canines are large, recurved, and pointed, being placed at a considerable distance from the incisors to accommodate the crown of the inferior canine. The three anterior premolars of the upper jaw are two-rooted, with the exception of the first, which is probably single-rooted. They have comparatively simple compressed crowns, with a principal cusp and a posterior basal lobe, surrounded by a basal cingulum. The fourth is more complex, and resembles the true molars posterior to it. Like them, it has three principal cusps, of which two are external and one internal, giving to the crown a triangular shape. In the first true molar the postero-external angle of the crown is produced into a strong blade-like process, a development of the cingulum which is conspicuous in all. The last molar of this series is bicuspid, the posterior of the two external cusps being absent.

In the lower jaw both the premolars and molars are remarkable for their simplicity. The first premolar is single-rooted, and has a subconic crown, as in the dog. The teeth behind it are two-rooted, and

418   DENTAL ANATOMY.

have a general premolariform appearance, the true molars exhibiting but little departure from the conical pattern of the lower Vertebrata. As in the premolars of the dog, their crowns are laterally compressed, of a

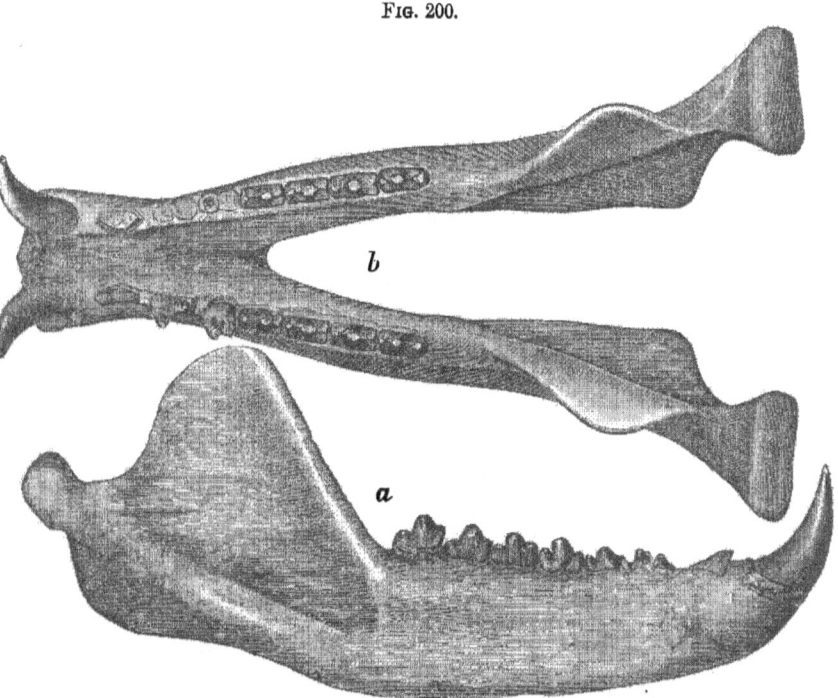

Fig. 200.

Mandible of *Mesonyx ossifragus*, Cope, from the Wasatch Epoch of the Big Horn River, Wyoming, one-third natural size (after Cope).

triangular form when viewed from the side, having a principal median cusp, to which are added an anterior and posterior smaller one from the cingulum.

It is a matter of considerable interest to find in this ancient representative of the unguiculate series so simple and generalized a dentition, inasmuch as it furnishes a key to an interpretation of the lobes and cusps of the teeth of many of the succeeding forms. It is more than probable that this particular species is not the original ancestral form from which the others have been derived, on account of certain characters presented by the skeleton, but, as far as the teeth of the lower jaw are concerned, they exhibit just such a transitional condition between the primitive cone of the theromorph Reptilia and the lowest forms of mammalian teeth as we would most reasonably expect to find in the primitive ancestor.

The various steps in this process of dental evolution I conceive to have been as follows: (1) additions to the anterior and posterior edges of the cone and the formation of a cingulum; (2) division of the single root into two; (3) addition of basal cusps from the cingulum. It is a fact worthy of notice that in the conical dentition the teeth of one series do not exactly oppose those of the other, but close in the intervals between them. This in animals that attempted to crush a morsel of food would cause stimulation of the anterior and posterior edges of the tooth, thereby determining the point of the greatest nutritive activity and consequent growth. Long-continued vertical pressure I believe to be an adequate cause for the appearance of the wrinkle or fold of the enamel covering at the base of the tooth which is designated as the cingulum.

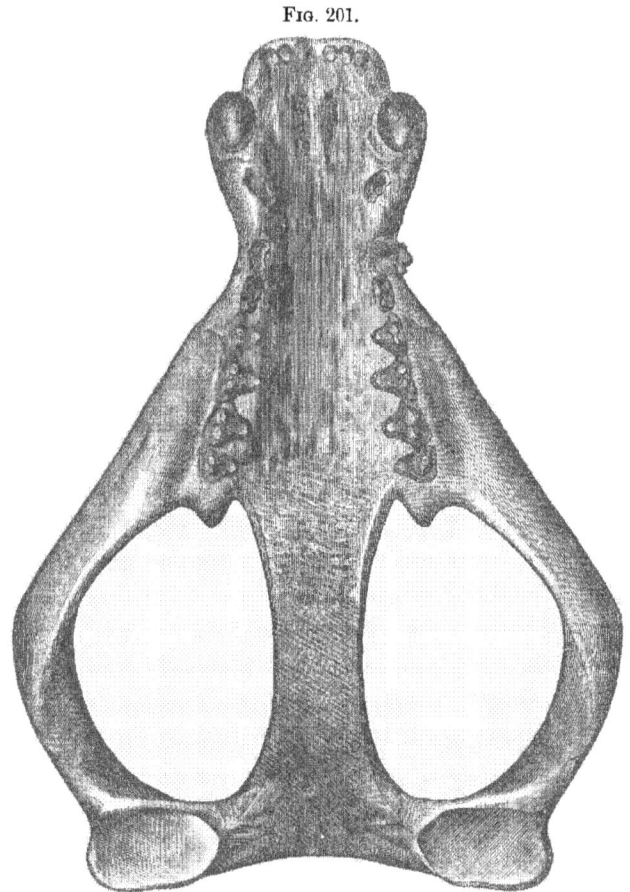

FIG. 201.

Skull of *Mesonyx ossifragus*, anterior to post-glenoid process, one-third natural size, from the Wasatch beds of Wyoming (after Cope).

The formation of two roots I believe to have been the result of the inequality of pressure exerted upon each tooth during the act of mastication, whereby there was an effort to displace the tooth in an antero-posterior direction, or, in other words, to give it a forward and backward rocking movement, as the greatest pressure was in front of or behind it. This would cause the stimulation of the anterior and posterior faces of the root, and as a consequence of this a vertical groove was first formed upon each side, which eventually coalesced, dividing the root into two. As we have already seen, this condition is found in a theromorph reptile, and is likewise to be found in the premolars of many existing animals. The development of basal cusps would naturally follow at those points where the crown sustained the greatest amount of resistance, which would be at the base of the triangle.

It is a rule of pretty general application in heterodont teeth that the molars are more modified than the premolars. This, in all probability, results from the greater mechanical advantage which is gained by bringing the morsel to be crushed or divided to the posterior part of the mouth; that is to say, the resistance as near to the power as possible. The power in this case is the muscles which close the mouth, which, being attached to the posterior part of the jaw, exert the greatest influence upon those teeth in the vicinity of their attachment.

The next step in dental complication is seen in the genus *Dissacus*, from the lowest Eocene, the lower teeth of which are represented in Fig. 202. They are very similar in general appearance to those of

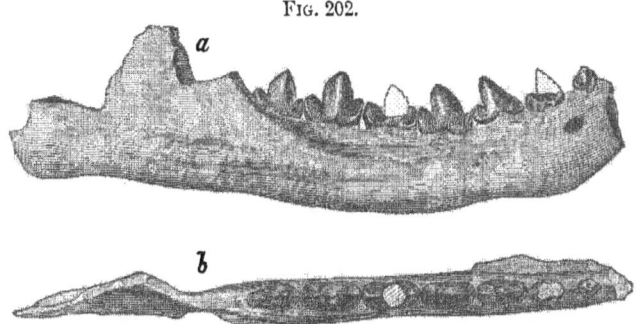

FIG. 202.

Dissacus navajovius, Cope. Right Mandibular Ramus, three-fourths natural size; *a*, external; *b*, superior view, from the Puerco Beds of New Mexico (after Cope).

Mesonyx, with which they also agree in number. There is, however, an additional cusp developed upon the inner side of the median cone near its summit, which is the homologue of the internal tubercle of the inferior sectorial of the dog, as well as that of many other animals of the unguiculate series. The upper teeth are not known. The genera in which the mandibular teeth present this premolariform structure are associated by Cope into a family which he calls the *Mesonychidæ*.

As a probable derivative of this family we have the extinct family *Hyænodontidæ*, of which the teeth of the single genus *Hyænodon* are represented in Fig. 203. This animal is known, so far, from the Miocene deposits of this country and Europe only, and has been shown by Prof. W. B. Scott of Princeton College to be a near relative of *Mesonyx*. The dental formula is, according to Gaudry, I. $\frac{3}{3}$, C. $\frac{1}{1}$, Pm. $\frac{4}{4}$, M. $\frac{2}{3}$ = 42. The incisors resemble those of the dog, the median pair being the smallest, the outer pair the largest. The canines are large, pointed, and recurved. The anterior premolars above are two-rooted and have premolariform crowns. The third and fourth are three-rooted, with

FIG. 203.

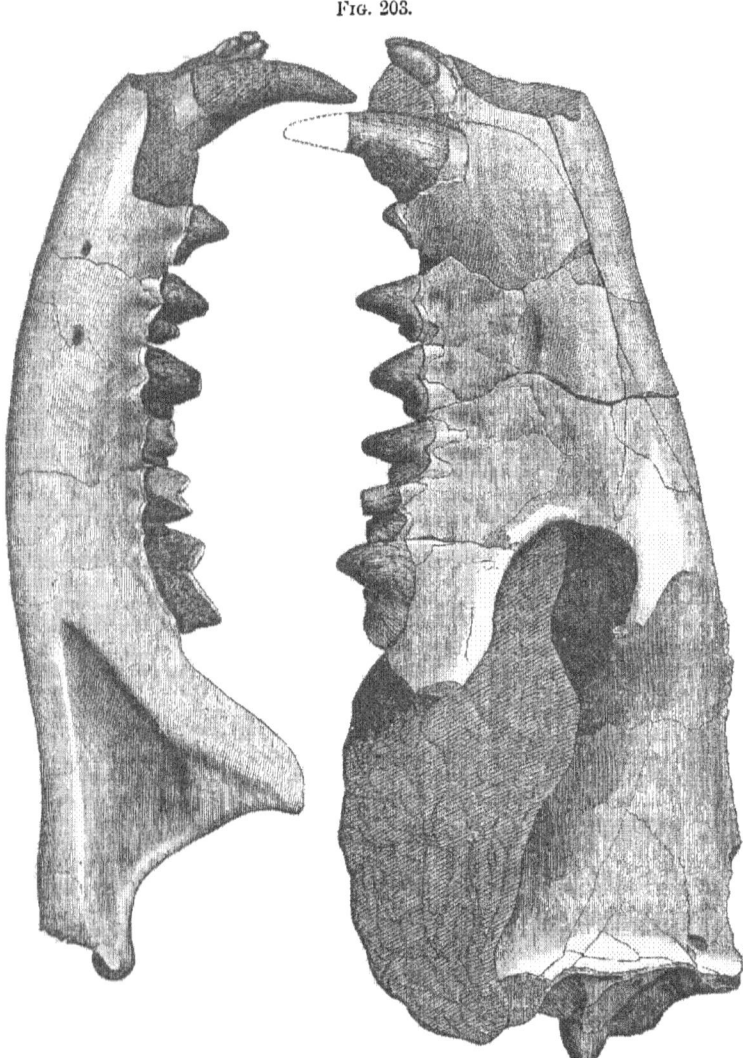

Hyænodon horridus, Leidy. Skull, one-half natural size (from Cope, after Leidy).

three external and one internal cusp, which in the third premolar is small and placed far back; in the fourth it is large and has a position nearer the middle of the tooth; while in the first and second molars it is anterior and more or less rudimental. The anterior cusp and median cone in these latter teeth form cutting blades and are truly sectorial in their nature.

In the lower jaw the first premolar is two-rooted, with a compressed crown without basal lobes. The second, third, and fourth have well-developed posterior basal lobes, with the anterior absent. The first true molar has three lobes, which form an imperfect trilobed sectorial blade. The second is also trilobed, having the anterior cusp and median cone developed into a true sectorial, while the posterior lobe forms a cutting heel. In the third the heel is rudimental or absent, and the anterior lobe and median cone are modified into a perfect trenchant blade.

From the *Mesonychidæ* we pass, through *Dissacus* and *Triisodon*, to another extinct family, which Prof. Cope calls the *Leptictidæ*. In an

FIG. 204.

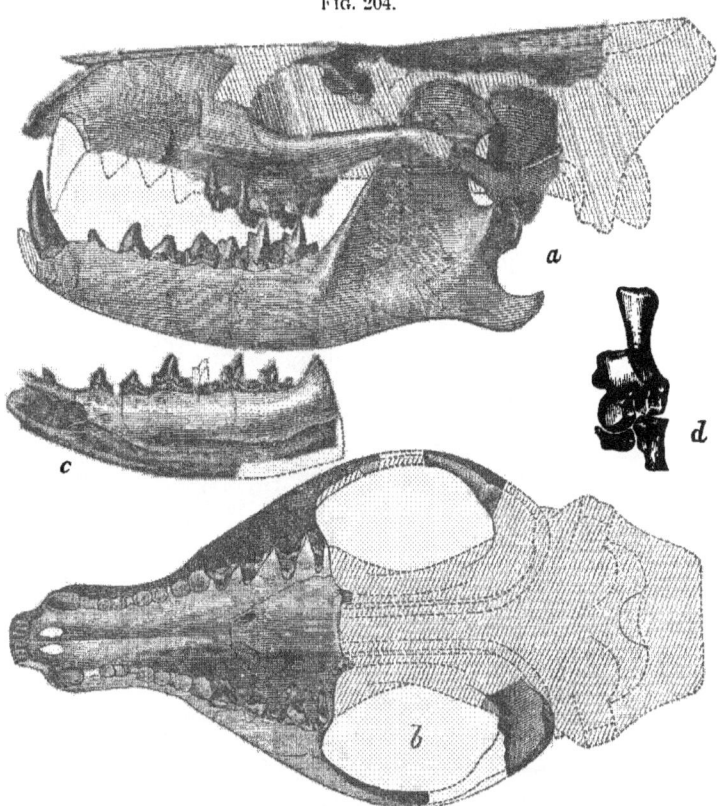

Skull and Part of the Posterior Foot of two individuals of *Stypolophus whitiæ*, Cope, two-thirds natural size: *a, b*, side and under views of the skull; *c*, portion of lower jaw; *d*, ankle-joint (after Cope).

Eocene genus of this family, *Stypolophus*, Cope (Fig. 204), the dental formula is I. $\frac{3}{3}$, C. $\frac{1}{1}$, Pm. $\frac{4}{4}$, M. $\frac{3}{3} = 44$. The incisors, canines, and premolars are very like the corresponding teeth in *Mesonyx*. The last superior premolar is, like the true molars, tritubercular, with the pos-

TEETH OF THE VERTEBRATA. 423

terior external angle produced into a prominent process, which is connected with the postero-external cusp by a sharp cutting ridge. The cingulum also furnishes a broad ledge upon the outside of the two external cusps.

In the three inferior true molars the median cone, the anterior basal lobe, and the internal tubercle are all well developed, and are disposed in such a manner as to form an equilateral triangle, with the apex directed forward and the base backward. Of these, the anterior basal lobe occupies a position at the apex of the triangle, the median cone and internal lobe being placed at the external and internal angles of the base respectively. The posterior basal lobe is also present in the form of a low heel, which may in some genera retain a simple cutting form or may be broken up into several and become " basin-shaped." This form of tooth Prof. Cope proposed some years ago to designate by the name of "*tuberculo-sectorial.*" There can be little doubt that it furnishes the point of departure for the sectorial teeth of the lower jaw

FIG. 205.

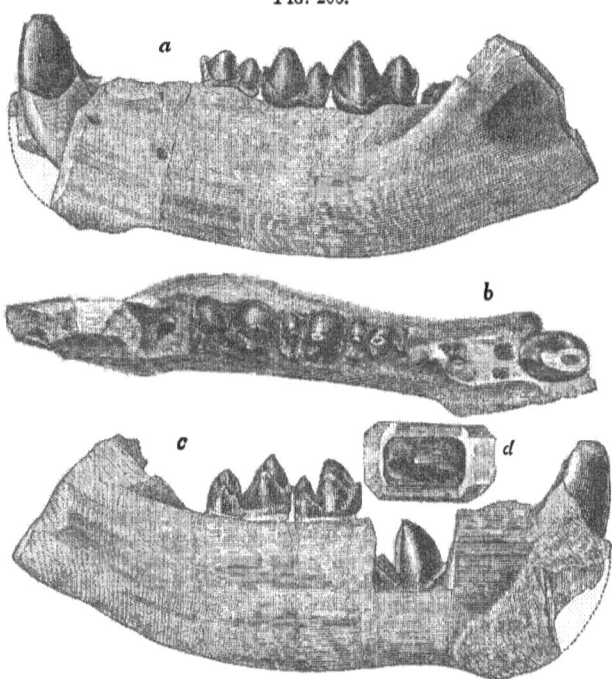

Left Mandibular Ramus of *Tritsodon quirirensis*, three-fourths natural size, from the Puerco of New Mexico; *a*, external view, displaying last temporary molar in place; *b*, the same from above; *c*, the same, internal side, the temporary molar removed and the permanent fourth premolar displayed in the jaw; *d*, the fourth premolar seen from above (after Cope).

of the modern *Carnivora* on the one hand, and the quadrituberoular lower molar of the entire unguiculate series on the other.

As will be seen, it displays the same elements that are found in the inferior sectorial of the dog, the only difference being in the relatively smaller size of the internal tubercle and the modification of the primitive cone and anterior basal lobe into a more perfect sectorial form in the dog. The quadrituberculur tooth has been derived from this by the suppression of the anterior basal lobe, the reduction in size of the median cone and internal tubercle, and the division of the heel into two cusps, whereby the median cone becomes the antero-external tubercle, the internal tubercle the antero-internal, and the heel the two posterior ones.

The genus *Triisodon* of Cope (Fig. 205) affords a perfect transition between *Stypolophus* and *Dissacus*, as far as the pattern of the inferior teeth is concerned. In another genus, *Chriacus*, Cope, from the Lower Eocene, which is provisionally referred to this family, the upper molars are tritubercular with a strong internal cingulum, which develops a rudimental fourth cusp behind. This forms one of the examples referred to in which it is difficult to say whether the teeth in question are tritubercular or quadritubercular, and goes far toward invalidating the definition of the *Creodonta* as given by Cope. This author says in reference to this family, and more especially to this genus: "Two groups are easily recognized among the *Leptictidæ*. In the first of these the last or fourth inferior premolar is a simple premolariform tooth, different from the inferior true molars and without any internal cusp. In the second division the fourth inferior premolar is either like the first true molars or approximates their form by the presence of an internal tubercle. To the latter group belongs the genus *Chriacus*, which, from the slight development of the fourth inferior premolar, approximates the first division. This genus may, however, be improperly referred to the Creodonta."[1]

Still another genus of this family, *Mioclænus*, also from the Eocene, presents truly quadritubercular lower molars. The premolars are simple and conical, and differ widely in their structure from the molars. The superior true molars are similar to those of the preceding genus, with the exception that the fourth tubercle is better defined and furnishes another example of the transition between the tritubercular and quadritubercular condition.

Fig. 206.

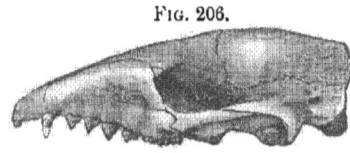
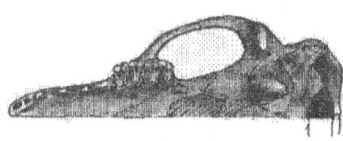
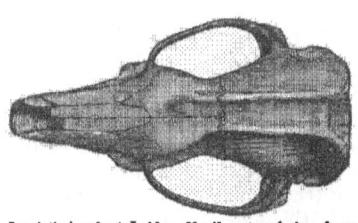

Leptictis haydeni, Leidy. Skull, natural size, from the White River beds of Nebraska (from Cope, after Leidy).

In the typical genus *Leptictis* of Leidy (Fig. 206) the dental formula is probably the same as that of *Stypolophus*, the lower jaw being imperfectly known. The upper teeth

[1] "The Creodonta," *American Naturalist*, April, 1884, p. 348.

are like those of *Stypolophus* in having the fourth premolar and all the true molars trituberular. There is no broad ledge external to the outer cusps, however, as in that genus, and the posterior external angle of the crown is not produced. It is from the White River Miocene of this country, as is also a nearly-related genus, *Ictops* of Cope. The only difference between these two genera is the more complex form of the fourth superior premolar in the latter.

The living genus *Centetes*, or tenrec of Madagascar, is closely related to this family, and differs only in the incomplete condition of the zygomatic arch. The number and form of the teeth are very like those of *Leptictis*, and it is highly probable, as Cope suggests, that *Centetes* is the living descendant of this genus.

Another quite remarkable genus which Cope places in this family is from the Eocene, and was described by him under the name of *Esthonyx*. Its dental formula (Fig. 207) is I. ½, C. ¼, Pm. ¾, M. ¾ = 34. The single superior incisor of each side is greatly enlarged, and exceeds the canine in size. The first premolar is small and has a simple crown. The next is larger, and is trituberular. The third is like the true molars, with the exception that it lacks the internal cingulum. The true molars have two external cusps, bordered upon the outside by a broad ledge which is produced anteriorly into a marginal cusp. There is a large internal cusp, from which is developed at its inner posterior extremity

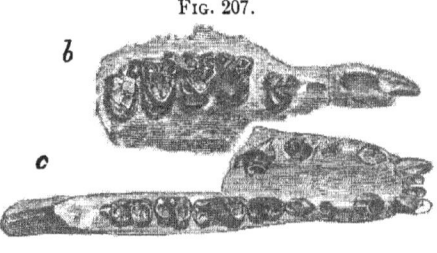
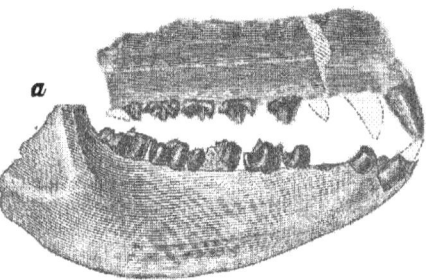

FIG. 207.

Esthonyx burmeisteri, Cope: *a, b, c*, parts of upper and lower jaws, two-thirds natural size (after Cope).

a strong cingulum, the representative of the fourth cusp, which is continued thence around the base of the crown behind to join its broad external portion. This is another case wherein the tritubercular and quadritubercular question is involved in uncertainty.

In the lower jaw the median incisors are small, the outer pair enlarged, almost equalling the canines. The first two premolars are small, the third larger, resembling the true molars somewhat in form. The molars support two Vs, of which the anterior is most elevated. An analysis of the crown shows it to be of a modified tuberculo-sectorial nature, wherein the three anterior cusps are connected by ridges that extend quite to their summit and form the anterior V. The broad heel displays two cusps connected by a strong ridge; from the outer of these, again, another ridge passes obliquely forward to join the internal cusp, thereby completing the second V. The last molar has a fifth cusp behind, in this respect resembling many of the lemuroids.

The family most nearly approximated to this genus is that including the shrews (*Soricidæ*), which always have two incisors both above and below, greatly enlarged. In *Blarina talpoides*, a living species of this country, the teeth (Figs. 208, 209) are thirty-two in number, of which

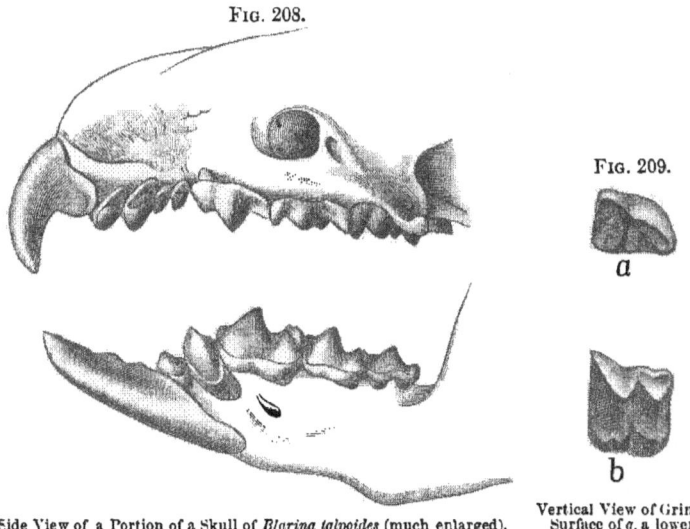

Fig. 208.

Side View of a Portion of a Skull of *Blarina talpoides* (much enlarged). The sixth tooth of the upper series is placed somewhat internal to the others, and is not represented in the drawing.

Fig. 209.

Vertical View of Grinding Surface of *a*, a lower molar, and *b*, an upper molar (enlarged.)

twenty belong to the upper and twelve to the lower jaw. Owing to the very early co-ossification of the premaxillæ with the maxillæ and the paucity of suitable material, I am at present unable to give the proper dental formula of this animal. Neither can I find any statement upon the subject further than that of Owen, in which he refers to the European species, *Sorex araneus*, with only eight teeth upon each side in the upper jaw, and says:[1] "The determination of the small teeth between the large anterior incisors and the multicuspid molars depends upon the extent of the early ankylosed intermaxillaries; the incisors being defined by their implantation in these bones, the succeeding small and simple crowned molars must be regarded as premolars, not any of them having the development or office of a canine tooth: their analogues in the lower jaw are implanted by two roots.". If he confines his statement to this species, it is probably correct to refer all those teeth between the large incisors and the anterior one of the last three to the premolar series; but in *Blarina* there are six teeth to be disposed of. Allowing the normal number of premolars (four), we have either three incisors and no canine, or, as is most probable, two incisors and a canine, which would give the following formula: I. $\frac{2}{2}$, C. $\frac{1}{0}$, Pm. $\frac{4}{2}$, M. $\frac{3}{3}$. This determination of course may prove to be incorrect.

[1] *Odontography*, p. 418.

The two large incisors above are hook-shaped, with a prominent posterior ledge at the base, which in some species of this family is produced into a strong basal cusp. The next two teeth are much smaller and subequal, while the three following, rapidly decrease in size, the last becoming very small. Exclusive of the large hook-shaped incisor, the above-mentioned teeth display a principal cone with an internal lobe and an external cingulum, which is most distinct in the fourth, fifth, and sixth. The next tooth, which I take to be the last premolar, marks an abrupt change both in the character and size of the teeth of the upper series. It is almost equal in size to the first true molar, which it resembles very closely in structure.

The crowns of the molars may be described as consisting of four principal cusps or tubercles, of which two are external and two internal, and are therefore quadrituberculan. The two external cusps stand at the apex of two Vs, which open externally, giving to this part of the tooth a distinct W pattern. At the extreme antero-external angle of the crown there is a considerable cingular cusp, which is connected with the main antero-external tubercle by a prominent ridge, thereby forming the first downward stroke of the W. From this main tubercle another well-marked ridge passes outward and backward to another small cusp of the cingulum, situated at a point midway between the two main external tubercles on the outer edge of the crown, forming the first upward stroke of the W and completing the first V. The second downward stroke of the W is furnished by a ridge connecting the small median marginal cusp with the postero-external tubercle, while the second upward stroke of the W is formed by a ridge continued outward and backward to the postero-external angle of the crown, where it terminates in as light enlargement.

From the apex of each of the Vs a high ridge passes inward, meeting at the antero-internal angle of the crown, forming a distinct U. The inner part of this crest is the antero-internal tubercle. The postero-internal tubercle stands just behind it, and exhibits a somewhat crescentiform pattern. It will be seen that the tooth just described does not differ materially from those of some other insectivores already noticed, especially *Esthonyx*. The principal differences are to be found in the greater development of the marginal cingular cusps and connecting ridges upon the external part of the crown, which we have, in a measure, foreshadowed in *Esthonyx* and others.

In the lower jaw the single pair of incisors are large, scalpriform, and procumbent. The two succeeding premolars are small, single-fanged, and have simple crowns. The first two true molars are the largest of the molar and premolar series, and exhibit a structure identical with that of *Esthonyx*. The last is very small, and corresponds with the last tooth of the upper jaw, which frequently disappears. The crowns of all the teeth are stained a deep wine-color by a pigment which penetrates the substance of the enamel, this tissue being remarkable for its thickness in all the Insectivores.

Considerable discussion has taken place in regard to the nature of the external Vs and the exact homology of the two external tubercles. Since this W-structure is common to the superior true molars of all the moles, shrews, and insectivorous bats, as well as some others of the unguiculate series, it is desirable to have a thorough understanding of it. Cope maintains[1] that the median and anterior marginal cusps are the homologues of the two external tubercles of the teeth of such a form as *Stypolophus* (Fig. 204), and that the two cusps, which are here homologized as the representatives of the two external tubercles of this genus, he proposes to call intermediate tubercles. Mivart, on the other hand, holds[2] that all the marginal cusps are developed from the cingulum, and that the true external tubercles have come to occupy a more and more internal position on the crown—a view which I believe to be correct.

The evidence upon which I base my opinion is to be found by examining the teeth of such genera as *Stypolophus*, *Esthonyx*, and *Scapanus* of the unguiculate series, and *Thylacinus*, *Didelphys*, *Phascogale*, and *Dasyurus* among the marsupials. In the genera *Stypolophus* and *Esthonyx*, as we have already seen, the way is paved, so to speak, for the formation of the two Vs by the appearance of a broad ledge external to the two main outer cusps and the elevation of the cingulum into a small cusp at the antero-external angle of the crown, as well as the backward prolongation of the postero-external angle and its connection with the postero-external tubercle by a strong ridge. This latter ridge I regard as the strict homologue of the last upward stroke of the W. In these two genera the only modifications necessary to produce the W would be greater separation of the two external tubercles and the presence of a median marginal cusp connected with them by ridges.

In the genus *Scapanus*, or hairy-tailed moles, of this country the fourth superior premolar does not exhibit the W-shaped arrangement in the same perfection that the true molars do, the anterior V being rudimental or absent. The cusp at the antero-external angle, however, is present, and can be clearly shown to be of a cingular origin. It cannot therefore, as Cope supposes, represent the true antero-external tubercle, in this tooth at least. In the second unworn true molar of *Dasyurus* (Fig. 210) all the marginal cusps are present, but the median one is not connected with the two main external tubercles by ridges, leaving the W imperfect. In this tooth nothing is more apparent than the cingular origin of *all* the marginal cusps; and no one can doubt, it appears to me, that they are strictly homologous with the cusps in a like position in the molars of those animals in which the W is perfectly formed.

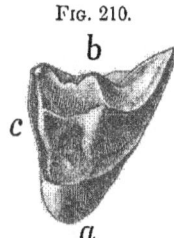

FIG. 210.

View of the Grinding Surface of an Unworn Molar Tooth of *Dasyurus* (enlarged): *a*, internal; *b*, external; *c*, anterior aspect of the crown.

A careful consideration of the teeth of the genera above mentioned in my judgment effectually disposes of the whole question, and demonstrates beyond doubt the correctness of the position here maintained, notwithstanding the conclusions of so high an authority as Prof. Cope to the contrary.

Galeopithecus, or the so-called flying lemur, constituting another

[1] "Mutual Relations of the Bunotherian Mammalia," *Proceed. Acad. Nat. Sciences Philadelphia*, 1883, pp. 81-83.
[2] *Journal of Anatomy and Physiology*, ii. 138, figures, 1868.

family (*Galeopithecidæ*), is quite aberrant in the form of its incisors and some of the premolars. The incisors are two in number upon each side in the upper jaw, and those of the opposite sides are separated by a wide edentulous space; the first is minute and comparatively simple; the second is relatively large, two-rooted, and in every way similar to the tooth behind it, which is lodged in the maxillary bone. The form of the crown is that of a greatly flattened cone with anterior and posterior cutting edges. The anterior edge is interrupted by one minor denticle, the posterior by four, making it distinctly serrated. The next tooth behind this one is sometimes called a canine, but it is more probably a premolar; if this be the case the premolars are three in the upper jaw. The last premolar is like the molars, with three principal cusps and two small intermediate ones.

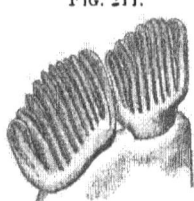

Fig. 211.

Two Incisors of the Lower Jaw of *Galeopithecus*, external view (enlarged).

In the lower jaw the incisors (Fig. 211) form a continuous arch around the alveolar margin, and are of a most remarkable pattern. They are four in all, of which the outer pair is somewhat the larger; they have broad incisive crowns, which are cleft to the base by deep vertical fissures like the teeth of a comb. In the middle pair there are seven such fissures, dividing off eight slender columns, whereas in the lateral pair there are ten.

The next tooth has a somewhat similar shape, but there are only four fissures, which do not penetrate so deeply; its crown cannot therefore be said to be more than serrate. The true molars are quintitubercular, with very elevated cusps.

In the European mole (*Talpa europea*), which may be taken as a fair representative of the family *Talpidæ*, the dental formula is I. $\frac{3}{3}$, C. $\frac{1}{1}$, Pm. $\frac{4}{4}$, M. $\frac{3}{3}$ = 44. The incisors of the upper series are normal both in size and structure. The upper canine is large, recurved, and pointed, and exhibits the remarkable peculiarity of being implanted by two fangs. The premolars are simple compressed teeth, increasing progressively in size from the first to the fourth. The true molars are tritubercular, with the W-shaped structure externally.

Fig. 212.

In the lower jaw the first four teeth are small and incisiform; the next is large, two-rooted, and caniniform, performing the function of the inferior canine. It is, however, really a premolar, since it closes behind the superior canine, and not in front of it. The tooth immediately in front of it is the true canine, notwithstanding its small size and incisive office. The three succeeding premolars are similar to the corresponding teeth above. The true molars are quadritubercular, or rather intermediate between the tuberculo-sectorial and the quadritubercular patterns.

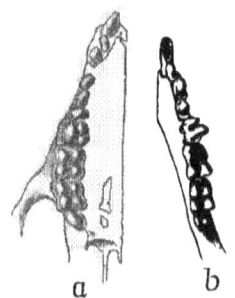

Vertical View of *a*, the upper jaw, and *b*, the lower jaw, of European hedge-hog (*Erinaceus*).

In the hedge-hogs (*Erinacidæ*) (Fig. 212) and the elephant shrews

(*Macroscelidæ*) the molars are quadritubercular both above and below, and exhibit no traces whatever of the complex W-structure. If it is imperative to make any division of the Insectivora upon the characters of the teeth, I would suggest that the W-arrangement of the cusps of the superior true molars be considered as available for the purpose, although I would be seriously disposed to question this character alone as indicative of community of descent.

Another family of the *Insectivora* which in all probability stands in ancestral relationship to the *Carnivora* is the one which Cope calls the *Miacidæ*. It is represented by two genera, *Didymictis* and *Miacis*, both from the Eocene of North America. In this family we have, as Cope remarks,[1] "the point of nearest approximation of the Creodonta and Carnivora. This is indicated by the fact that the sectorials are sectorials both by position and form, such as are not elsewhere met with in the Creodonta. The genera might readily be taken for members of the Canidæ and Viverridæ (dogs and civets) but for the structure of the astragalus, which is thoroughly creodont."

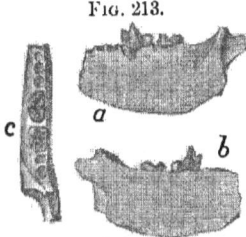

Fig. 213.

Fragment of the Lower Jaw of a species of *Miacis*: *a*, external, *b*, internal, and *c*, vertical views.

The genus *Didymictis* may be certainly regarded as the ancestor of the civets, while it is more than probable that *Miacis* (Fig. 213) was the immediate progenitor of the dogs. In the teeth of *Didymictis* (Fig. 214) the dental formula is not completely

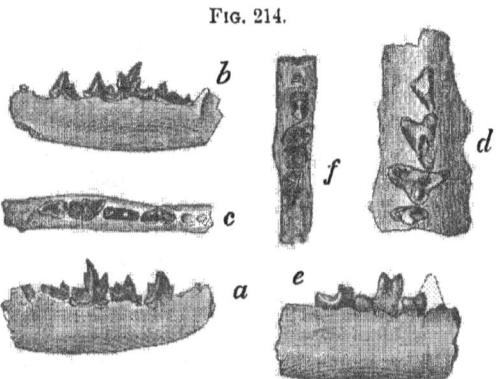

Fig. 214.

Two Species of *Didymictis*: *a*, *b*, *c*, internal, vertical, and external views of lower jaw of *D. dawkinsianus*, from Big Horn Beds; *d*, *e*, *f*, *D. haydenianus*; *d*, upper-jaw fragment, vertical view; *e*, fragment of left ramus, inner side; *f*, vertical view of same,—all natural size (after Cope).

known, but most probably it is I. $\frac{3}{3}$, C. $\frac{1}{1}$, Pm. $\frac{4}{4}$, M. $\frac{2}{2} = 40$. The fourth premolar above is sectorial in form, the two true molars tubercular.

In the lower jaw the first true molar has lost much of its typical tuberculo-sectorial structure, which is best seen in the decreased size of the internal tubercle and the tendency of the anterior basal lobe and the

---

[1] "The Creodonta," *American Naturalist*, May, 1884, p. 483.

primitive cone to fuse into a cutting blade. A single tubercular molar follows the sectorial, which exhibits, as does the second lower true molar in the dog, a reduced or degraded condition of the tuberculo-sectorial pattern.

Miacis has three true molars in the lower jaw, of which the first is sectorial, in this respect resembling very closely the lower jaw of the dog; its complete dental formula is not known.

TEETH OF THE PROSIMIÆ.—With this group we enter that division of the *Bunotheria* which leads out to the monkeys and man. Its palæontological history reveals an antiquity quite equal to that of any other of the Monodelphia, continuing backward to the lowest Eocene. It is customary with most naturalists to regard the *Prosimiæ* as widely separated from the *Insectivora* on account of the higher order of brain-structure which the living representatives of the lemurs display, and they are accordingly placed near the Primates. Owing to the perishable condition and non-preservation of the soft parts in the extinct forms generally, we will never be able to know the exact structure of their brains, but must be content to judge of its generalized or specialized character by the mould of the cranial cavity, which in many respects is unsatisfactory.

It can be shown in the ungulate series that the lowest Eocene representatives possessed brains, judging from the cranial casts, almost as low in the scale of organization as that of the lowest known mammals, and it is likewise true that the brains of the Eocene prosimians were more generalized than those now living. I do not think there can be any radical differences shown to exist between the structure of the brain of such forms as the squirrel shrews (*Taupaiadæ*), the elephant shrews (*Macroscelidæ*), and the *Galeopithecidæ* of the insectivores, and the true lemurs (*Lemuridæ*), the fossil *Adapis*, and others of the *Prosimiæ*. The very fact of their remote antiquity and appearance in an age when the brain-development of all the Mammalia was small would of itself lead to the supposition that they too at first possessed brains of lowly organization. It should be here stated that very few skulls of the Eocene prosimians are known.

The dental formula of the spectrum lemur (*Tarsius spectrum*) is I. $\frac{2}{1}$, C. $\frac{1}{1}$, Pm. $\frac{3}{3}$, M. $\frac{3}{3}$. Of the two pairs of incisors in the upper jaw, the median is much the larger; they are closely approximated, long, pointed, and conical, and are surrounded at the base by a prominent cingulum, which is well defined upon the anterior face of the crown. The next pair are much smaller, and also have pointed crowns and basal cingula. The upper canines are about equal in size to the median incisors, which they resemble both in the form of their crowns and the cingulum at the base.

The first premolar is the smallest of the three, and is placed just behind the canine; its crown is simple and pointed. The next two are larger and imperfectly two-lobed, the internal lobe being represented by a strongly-developed cingulum which continues around upon the outer face of the tooth. The true molars are subequal in size and tritubercular. The two external cusps are well developed, and placed at the external border of the crown. The internal lobe is relatively large, and occupies a position opposite the interval of the two external. A moderate cingulum is developed on its internal aspect, and continues around to the outside of the tooth.

In the lower jaw the single pair of incisors come close together above the symphysis, and completely fill the space between canines; they have conic crowns and are smaller than the median pair above. The canines are larger than those above, and like them have pointed, slightly recurved apices. The premolars resemble those above, with the exception that the internal lobe is absent. The true molars are quadritubercular, with the two anterior slightly elevated. A trace of the anterior basal lobe is visible, and is best marked in the first, which brings the structure of the tooth into close correspondence with the tuberculo-sectorial of the Insectivores, and strongly suggests its derivation from it, as so many other examples of a similar kind do. The last molar displays a fifth lobe behind the two posterior ones, and is therefore quintitubercular. This species is the sole representative of the family *Tarsiidæ*.

In the typical lemurs of the family *Lemuridæ* the two pairs of upper incisors are separated from each other by a wide space in the centre, both being small and subequal. The superior canines are large; the premolars, with the exception of the first, have a small internal cusp. The molars are quadrituberculous by reason of the internal cingulum rising up into a cusp at the postero-external angle of the crown. Various intermediate conditions between the perfect development of this cusp and its almost complete absence are to be seen. The fourth premolar, too, is in some genera like the true molars, in which case the last molar is small.

The incisors of the lower jaw are two in number upon each side, and are long, slender, laterally compressed teeth, having a procumbent implantation. The canines resemble them very much both in shape and position, being a little larger. The first premolar is large and caniniform, and would be readily taken for the canine at the first glance. The two following are smaller, and usually have simple crowns. The molars are truly quadrituberculous, the anterior basal lobe being entirely absent. The last molar may or may not have a fifth posterior tubercle.

TEETH OF THE TILLODONTA, TÆNIODONTA, AND DAUBENTONIOIDEA.—The aye-aye (*Chiromys*) of Madagascar is generally associated with the lemurs in the sub-order *Prosimiæ*, but naturalists—notably Profs. Cope and Gill—have seen fit to give it a rank equal to that of the lemuroids and place it in a distinct sub-order, *Daubentonioidea*, on account of the aberrant character of its teeth as compared with the lemurs. There are two other Eocene groups which go with it and constitute the first division of the order *Bunotheria*, according to Cope.

The dental formula of the adult aye-aye is, according to Owen, I. $\frac{1}{1}$, C. $\frac{0}{0}$, P. $\frac{1}{1}$, M. $\frac{3}{3}$ = 18. The upper incisors are curved as in the *Rodentia*, and deeply implanted in the jaw. Their exposed portions are contiguous, their widely-excavated fangs diverging as they proceed backward. The incisors of the lower jaw are similar in shape to the upper ones, and are implanted as far back as the coronoid process. They are all covered with enamel, both in front and behind, and grow from persistent pulps. In the entire investment of enamel they offer an important difference from the incisors of the rodents, which they otherwise closely resemble. The enamel being thicker upon the anterior than upon the posterior face of the tooth causes them to wear into chisel-shaped extremities, whereby the same effective gnawing instruments are produced as in the typical gnawing quadrupeds. The molar and premolar teeth are four in number upon each side in the upper, and three upon each side in the lower, jaw. They are implanted after a considerable interval behind the incisors, leaving a wide space or diastema, as in the *Rodentia*. The first and last molars of the upper series are the smallest, and have single roots; the second and third larger, and implanted by three fangs each. Their crowns have simple subelliptical grinding surfaces. The molars of the lower jaw are similar, the first being implanted by two roots, the second and third by one each.

The deciduous or milk dentition is I. $\frac{2}{1}$, C. $\frac{1}{0}$, Pm. $\frac{1}{1} = 12$. In the milk set a small incisor appears upon each side of the median pair, and is not replaced by a permanent one. Two teeth in this set occupy the spaces between the premolars and incisors above, and have been considered canines, they having no permanent successors. The single deciduous molar in each jaw is succeeded by the permanent premolar.

The *Tillodonta* is a group which was discovered and described by Prof. O. C. Marsh from the Upper Eocene deposits of Wyoming Territory. In the typical genus, *Tillotherium*, the dental formula, as given by this author is, I. $\frac{2}{2}$, C. $\frac{1}{1}$, Pm. $\frac{3}{3}$, M. $\frac{3}{3} = 34$. The median pair of incisors in each jaw are large and scalpriform, being faced with enamel, as in the rodents. They grew from persistent pulps, as is indicated by the large pulp-cavities at the base. The outer pair are small and did not grow persistently. The canines are much reduced, and placed well back in the alveolar border. The first of the three premolars of the upper jaw is small and simple, the other two being larger and of a more complex pattern.

The structure of the crowns of the superior molars is not very different from that of *Esthonyx*. Two external cusps are present, which are not well separated from each other; external to them is a broad cingular portion which is produced anteriorly into a process more marked than in *Esthonyx*. Internally two cusps are present, the posterior being lunate and consisting of a highly-developed cingulum. In the anterior, or that which corresponds with the antero-internal cusp of the quadrituberular molar, two well-developed ridges pass outward from its summit, one toward the anterior, and the other toward the posterior external angle of the crown, giving it a rounded U-shaped appearance. The pattern of the lower molars is identical with that of *Esthonyx*.[1]

In a general survey of the dentition of this genus I am compelled to dissent from the views expressed by Mr. Tomes, wherein he says that the molar teeth are of the ungulate type, and that the order combines characters of the Carnivora, Ungulata, and Rodentia. While it is true that the scalpriform incisors faced with enamel is a condition exhibited by the rodents, a condition also found in the *Toxodontia*, I fail to

[1] See Professor Marsh's monograph of this group, *American Journal of Science and Arts*, vol. xi., 1876, p. 249.

discover the faintest trace of either carnivorous or ungulate relationship. On the other hand, it seems to me that the evidence points strongly to the fact that this group is the direct descendant of *Esthonyx*, which preceded it in time. This is especially seen in the increased size of the mesial pair of incisors, the reduction of the canines, loss of one premolar in the upper jaw, and the remarkable similarity in the pattern of the molar teeth. That *Esthonyx* is an insectivore allied to the shrews there is scarcely any doubt. It is also probable that this group gave origin to the toxodonts, but the exact connections between them are not now apparent.

Another family, *Stylinodontidæ* of Prof. Marsh, makes approaches in this direction in the growth of the molars as well as the incisors from persistent pulps.

In the *Tæniodontia* the incisors are large and scalpriform, and were of persistent growth; the molar and premolar series are not separated from them by any diastema, in the lower jaw at least, and the canines, or those teeth regarded as such, in the inferior set also grew from persistent pulps, and have grinding crowns.[1]

TEETH OF THE PRIMATES OR QUADRUMANA.—The teeth of this order are closely affiliated with those of the typical lemuroids in the structure of the molars, and when compared with that of the other groups the amount of dental variation is comparatively insignificant. The order is naturally divisible into five families, of which the marmosets and platyrrhines of South America and the catarrhines and anthropoids of the Old World, as well as man, constitute the respective divisions. Of these families, the marmosets (*Hapalidæ*) are the most generalized and approach nearest to the lemurs in several important characters, prominent among which are the relatively smooth cerebral hemispheres, want of opposability of the thumb and its termination by a distinct claw instead of a nail,[2] and the possession of trituberoular instead of quadritubercular molars. Since they are found only in the New World, and as lemuroids were very abundant in this country in the Eocene Period, it seems probable that they are the derivatives of some member of this group. It is a fact worthy of notice that in the curious Eocene genus *Anaptomorphus* we have a near approach to the anthropoid condition of the teeth. In the shortness of the jaw and certain cranial peculiarities it also resembles the higher monkeys. For this reason Cope believes that the simians have descended directly from this lemur.

The dental formula of the genus *Midas* is I. $\frac{2}{2}$, C. $\frac{1}{1}$, Pm. $\frac{3}{3}$, M. $\frac{2}{2}$ = 32, which obtains in the one other living genus. The upper incisors have longitudinally flattened incisive crowns, with a prominent internal ledge at the base. The median pair is the larger, as is gener-

---

[1] Prof. Cope has established this sub-order upon the peculiar condition of the canine teeth of the lower jaw, or at least those which he supposes to be such; the only knowledge we have of the teeth of the upper jaw is confined to the large scalpriform incisors. This sub-order must be regarded as provisional until we know more of the upper teeth, as well as the relationship of some genera apparently intermediate between it and the *Tillodonta*.

[2] Many of the lemurs are provided with an opposable thumb, which is terminated by a distinct nail. In this respect the marmosets are even below the lemurs.

ally the case in all the Primates, and are in contact in the median line. The smaller outer incisors follow closely in the dentigerous border of the premaxillaries, after which there is a wide space, almost equal to the width of the two incisors, for the passage of the lower canine. The canines of the upper jaw are comparatively strong for the monkeys, and have pointed, slightly recurved crowns which project far above the level of the other teeth; there is a deep groove upon their anterior faces.

The premolars or bicuspids are three in number, and completely fill the interval between the canines and molars. The first is the smallest, and has a prominent pointed external cusp on the grinding surface, to which the cingulum adds a low U-shaped internal portion; the second and third are similar, except that the internal lobe is no longer cingular, the cingulum furnishing a second internal ledge. The true molars are two in number upon each side, in this respect differing from all known Primates. The only approach to this condition to be met with elsewhere in the order is in the dentition of man, in whom it appears, as we will hereafter see, that the last molar, or the "wisdom tooth," is gradually becoming rudimentary or defective in the higher races. Various causes have been assigned in explanation of this fact, one of which is that the greater development of the brain necessitates the expenditure of smaller amount of growth-force upon the maxillary bones, whereby insufficient room is allowed and the tooth stunted. If this be the real cause, it is difficult to understand why in the lowest representatives of the order—and those, too, in which the cerebral hemispheres are proportionally the smallest—the complete suppression of the last molar should have occurred.

The two pairs of lower incisors are small and of the usual incisiform pattern, being considerably smaller than the canines. The lower incisors of the allied genus, *Hapale*, are proclivous, the canines being relatively small and approximated to them, as in the lemurs, although not to so great an extent. The canines are almost equal to the upper ones in size, and follow the outer incisors without interruption. The three lower premolars are subequal, the summit of the first being elevated above the level of the succeeding teeth. In the first the anterior basal lobe, the principal cone, and an imperfect heel can be indistinctly made out, while in the second and third the internal tubercle is present. In the true molars there are four indistinct cusps; the anterior basal lobe has almost completely disappeared, and all the cusps are of equal height. A careful study of unworn teeth will show them to be a still further modification of the tuberculo-sectorial type, whereby the perfect quadritubercular has been produced.

The next division, Platyrrhines, or flat-nosed monkeys, constitute the family *Cebidæ*, in which the dental formula is I. $\frac{2}{2}$, C. $\frac{1}{1}$, Pm. $\frac{3}{3}$, M. $\frac{3}{3}$ = 36. They belong to the continent of South America, and have prehensile tails and generally rudimentary thumbs. The canines are usually strong and prominent, and the superior molars have a well-defined ridge connecting the antero-internal with the postero-external cusps, a remnant of the tritubercular condition. This ridge is found with varying constancy in the superior molars of all the Primates, and marks the connection between the internal cusp and the postero-external tubercle, which generally exists in the tritubercular tooth. The postero-internal cusp, which lies inside and behind this ridge, is the last one which has been added to complete the quadritubercular tooth in the upper jaw. In the squirrel monkeys of this family the lower incisors have a tendency to be proclivous, as in *Hapale* of the marmosets, thus retaining the lemurine character of these parts. No fossil remains of this family are known except from very late geological time, and these do not differ materially from those now living.

The teeth of the Catarrhines (*Semnopithecidæ*) show a reduction in the number of premolars, whereby the formula I. $\frac{2}{2}$, C. $\frac{1}{1}$, Pm. $\frac{2}{2}$, M. $\frac{3}{3}$ = 32, the same as that of man, is reached. The incisors are of the same shape as in man, the central pair being considerably larger than the outer pair. The canines are always strong and powerful teeth, and their apices are always elevated above the other teeth. They reach their maximum of development in the baboons, more especially in the dog-headed baboon, *Cynocephalus*, in which they are deeply grooved anteriorly. In this group the first premolar below is implanted by a double fang, with its apex directed upward and backward. The anterior root is naked for some distance, and presents in front a blunt edge which bites against the posterior edge of the powerful superior canine, giving to this part of the jaw a peculiar and characteristic appearance. The second lower premolar of *Cynocephalus* is quadritubercular, with all the cusps well developed, but in the macaques the posterior tubercles are not well defined. Both the upper and lower true molars increase in size from the first to the last, the last lower one being distinctly five-lobed.

In the semnopithiques the incisors are more nearly equal in size; the canines are smaller and less deeply grooved than in the baboons; the first and second molars are subequal, while the last lower molar is proportionally narrower, but still retains the fifth lobe. The typical cercopithiques have the last lower molar quadritubercular and all the molars subequal. Fossil remains of this family are known from the Miocene and Pliocene deposits of Europe and Asia, but no characters of unusual importance occur in their dentition.

The next family of this order includes the anthropoid or tailless apes, which are also confined to the tropics of the Old World. They constitute the family *Simiidæ*, and are distinguished from the preceding family, *Cercopithecidæ*, principally by the absence of the tail; from the succeeding family, *Hominidæ*, by the circumstance that the hallux is opposable, whereas in the latter it is in a line with the other digits and is not opposable. Other characters of considerable anatomical importance are also found which distinguish them from man.

The teeth of this family are the same in number as those of man, but considerable differences are found in the relative size of the canines and the last molar when compared with that which obtains in the human subject. Although they are organized substantially upon the same plan, the teeth are larger and stronger than in man. The orang (*Simia satyrus*) is probably the most human-like in its dentition, although in other respects the gorilla and chimpanzee most resemble man. The molar teeth in this animal are remarkable for the straight line in which they are implanted in both jaws, and contrast with the graceful curve they pursue in the normal human mouth.

The median pair of incisors are larger both above and below; in the upper jaw they are more than twice the size of the lateral pair, while in the lower jaw they are more nearly equal. Between the lateral pair and the canine above there is a considerable space, into which the lower canine bites. The canines are relatively large, and their apices rise far above the level of the surrounding teeth. They are imperfectly trihedral in form, with a trenchant edge behind. These teeth are larger in the male than in the female.

The premolars or bicuspids differ from those of man in the upper jaw in being implanted by three roots like the molars; their crowns are very similar to those of man, presenting essentially the same elements. The pattern of the crowns of the molars is like that of the human subject both above and below, but the last molar is as large as the others; it is implanted by three roots, and is always perfectly formed. In the lower jaw the two posterior molars slightly exceed the first in size, and the last is distinctly five-lobed. The first lower premolar is two-rooted, and has a faint resemblance to the corresponding tooth in the baboons. The second is also implanted by two roots, and its crown agrees with that of man.

In the other genera minor differences only are to be met with in the form, pattern, and arrangement of these organs.

The Human Dentition.

In this connection we come next to consider the teeth of man; and before so doing I am constrained to make some general remarks in regard to the position he occupies in the zoological scale. While it is undeniable that by virtue of his superior brain-capacity and intellectual development man is to be accorded a place at the head of the animal kingdom, it is nevertheless true that much of his anatomical structure has not been specialized beyond that of many of the lower forms. The fact that different members of the mammalian sub-class have been modified in different directions, some to fit one environment and some another, has led to the specialization of different sets of organs, and that, moreover, in different ways as the surrounding conditions and particular exigencies of the case have required.

It is these differences which enable the naturalist to construct zoological definitions of the major or minor groups, such as orders, sub-orders, families, genera, etc. The impracticability of determining which animal is highest or lowest in the scale of organization is thus rendered apparent from the fact that a comparison of different sets of organs is involved. Thus, in their dental, digestive, and limb structure the ungulates surpass all other Mammalia in complexity and specialization, and in these respects may be said to be highest, while in the matter of brain-development they are much inferior to others. The monkey line or Primates, on the other hand, of which man is at the head, retain a comparatively generalized structure of the limbs, teeth, and digestive organs, but have outstripped all others in the development of the cerebral nervous system.

It is only upon an evolutionary basis that we are enabled to comprehend the significance and import of the manifold modifications with which the morphologist is called upon to deal, and it is not at all unnatural that in the consideration of the human or any other dentition the student should first of all bend his energies to the discovery of the relative position which his subject holds in the system. All the evidence which anatomical and palæontological science can now bring to bear on the question tends to show that man is the legitimate product and highest expression of the evolutionary forces in that line of development which began with the Eocene lemuroids, however objectionable this conclusion may be to many. No adequate conception of his place in nature or the structure of any set of his organs can be had without a comparison with the other members of the stem to which he naturally belongs. This reason alone has induced me, somewhat contrary to custom, to give an account of the human dentition in this situation, rather than at the latter part of the present article.

Looked at from the point of view of the comparative odontologist, these organs present little of general morphological interest beyond that displayed by other Primates; but from the practical standpoint of the operative dentist they are of the greatest importance. In the course of this account many questions in connection with this latter phase of the subject will doubtless suggest themselves to the reader which are not within the scope of the present part of the work to discuss, its object being merely to outline the anatomy.

The dental formula of the human subject is, normally, I. $\frac{2}{2}$, C. $\frac{1}{1}$, Pm. $\frac{2}{2}$, M. $\frac{3}{3} = \frac{16}{16} = 32$, the same as that found in the Old World monkeys. Much variation from this number exists, however, by reason of the failure of development of the superior lateral incisors and of the third molars, the wisdom teeth or *dentes sapientiæ;* these molars may be present in the upper or lower jaw only, or they may fail to develop on one side in one or both jaws, or, again, they may be completely aborted. These variations are most frequently met with in the higher races of mankind, and are said to be of rare occurrence in the inferior races. The teeth are implanted in the alveolar process in such a manner in both jaws as to describe a regular parabolic curve, being uninterrupted at any point by the intervention of diastemata or spaces.

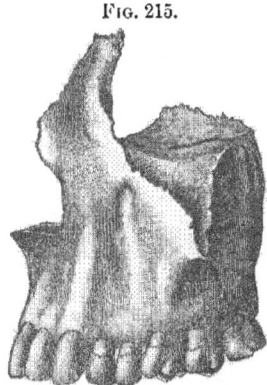

Fig. 215.

Superior Maxillary Bone of Man.

The summits of the crowns have, when normally developed, approximately the same level, the canines not excepted, thereby affording a marked contrast with the apes and monkeys, in which the crowns of the canines are always more elevated than the other teeth.

The incisors are four in number in each jaw, those of the upper being implanted in the premaxillary bones, which at an early period coalesce with the maxillaries. Of these, the central pair is the larger and has a slightly more anterior position than the lateral ones, on account of the curve of the alveolar border. Their incisive nature is manifested by the possession of a crown, which is bevelled on its palatine or lingual surface[1] to a cutting edge, being broader at the extremity than at the

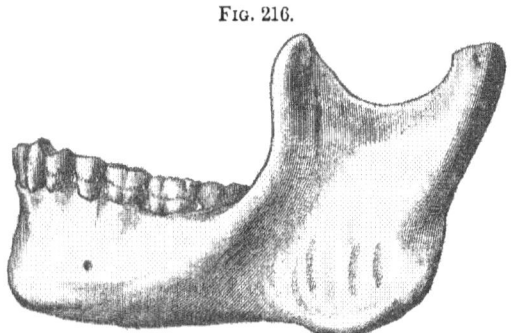

Fig. 216.

Inferior Maxillary Bone of Man.

base. The adjacent teeth are in contact at their coronal extremities, but on account of the narrower base a slight interval appears between them at the margin of the gum. The root joins the crown without any marked constriction, so that a neck can scarcely be said to exist; from this point it tapers gradually to an obtuse termination, being imperfectly trihedral in form and slightly recurved.

In newly-erupted teeth the cutting edge of the crown is divided into three inconspicuous cusps, which soon disappear through wear, leaving it smooth. The basal termination of the crown is indicated by the limit of the enamel covering, which is of greater vertical depth on the labial and palatine or lingual than on the lateral faces, so that if a line be drawn around the tooth at the most extreme basal portion of the enamel, it will touch only the labial and palatine prolongations, and not mark its exact limit on the mesial and distal surfaces. These

---

[1] The nomenclature of the various surfaces of a tooth as it stands in position in the jaw, it seems to me, is simplified by employing terms with the following signification: If the tooth-line were straightened out upon each side, the surface which looks away from the condyle would be *anterior*, and that which is directed toward it would be *posterior*; the surface directed toward the median line of the mouth would be *internal*, and that directed away from it *external*. In this system some confusion may arise with respect to the incisors and canines, in which the anterior surface is internal and conversely, owing to the curvature of the tooth-line; but while it has appeared to me best to speak of the surfaces as if the tooth-line were straight, I have in this paper adopted terms now most familiar to the dental profession, which are represented by the following: The surface looking toward the anterior part of the mouth and median line is called the *mesial surface*; its opposite, looking toward the condyle, the *distal surface*. In the superior row the surface which has been designated the internal I shall term the *palatal*, and in the inferior row the *lingual*, while the external surface is the buccal for the molars and bicuspids, and labial for the incisors and cuspids or canines. The triturating surfaces of the molars and bicuspids are termed *the masticating surfaces*, while the incisive surfaces of the incisors and cuspids or canines are denominated the *cutting edges*.

projections of the enamel present convex outlines basally, and are separated from each other by two wide V-shaped notches occupying the mesial and distal faces.

The labial aspect of the crown is convex from side to side, as well as from above downward, and is of greater vertical than transverse extent. Upon either side the crown is triangular in form, with the apex of the triangle terminating at each free angle of the cutting extremity, and the base directed toward the root; the basal part of the triangle is interrupted by the V-shaped notch already alluded to. That lateral surface which is directed toward the median line (mesial) is comparatively flat and most produced at the extremity, while the one which looks away from the median line (distal) is more rounded, having its terminal angle less produced. The interior or palatine surface is also triangular, but the base is formed by the free cutting edge and the apex turned toward the root. Usually, this surface is nearly flat, but in some examples it presents a broad central concavity whose depth may be considerably augmented by the presence of two marginal ridges meeting at the radicular extremity or apex of the triangle. These ridges, which are homologous with the cingulum of other teeth, sometimes develop a small cusp at their point of junction, in front of which there is usually a deep pit in the enamel—"a favorite site for caries." As a general rule, the cingulum is but faintly marked, and the posterior or palatine face is slightly concave.

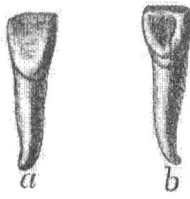

FIG. 217.

A Left Upper Central Incisor of Man: *a*, external or labial aspect; *b*, internal or lingual aspect.

The lateral incisors of the upper jaw are smaller than the median pair, but have approximately the same form. The labial face is more convex from side to side, and the outer or distal angle of the cutting edge is much more rounded off than in the median. The lingual surface may be slightly concave from above downward, and convex in the opposite direction, without any trace of the cingulum, or, as is most generally the case, it is concave, with the cingulum present, and elevated into a small cusp at the point of junction of the two lateral ridges. The basilar contour of the enamel covering is the same as in the preceding tooth. The root is more compressed laterally, of relatively greater length, and tapers more gradually to its termination, giving to the tooth a more slender and less robust appearance.

FIG. 218.

A Lower Incisor of Man: *a*, anterior, and *b*, lateral view.

The pulp-cavities of these two teeth have substantially the same shape, and the description of one will answer for that of both. Its form is that of an elongated tube, gradually increasing in diameter from the apical foramen in the apex of the root to a point which nearly coincides with the summit of the V-shaped notch in the enamel on the lateral surface of the crown, where it becomes contracted in an antero-posterior direction, but enlarged in its transverse diameter. It is prolonged upon either side into a slight cornua, which reaches but a short distance beyond the level of the general cavity; the one which corresponds to the internal or mesial angle of the cutting edge of the crown is a little the longer of the two.

The two pairs of lower incisors reverse the condition of the superior set, in that the central ones are the smallest. Their crowns have substantially the same pattern as those in the upper jaw, with the exception that an internal or lingual cingulum is never developed. They are readily distinguished from those above by their smaller size and greater lateral flattening of the roots. The pulp-cavity and basal enamel contour are like the corresponding teeth above.

The cuspids, canines, or "eye teeth," are the next in order behind the incisors; in both jaws they completely fill the gap between these latter teeth and the bicuspids, being in contact with them at the mesial and distal extremity of the crown. They are in every way stronger and more robust than the incisors, and are implanted by roots whose length, proportionate to that of the crown, is much greater. In the upper jaw these are indicated on the external surface of the maxillary bone by a vertical ridge or swelling which in many skulls extends quite as high as the lower border of the anterior nares.

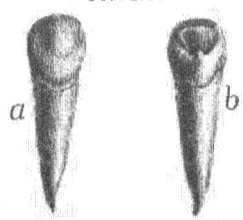

FIG. 219.

A Left Superior Human Canine; a, external, and b, internal view.

The crown is terminated by an obtuse point, which has a position in a line with the longitudinal axis of the root. Upon either side of this cusp the terminal extremity slopes away, but still retains a blunt cutting edge. When the median cusp is reduced by wear the crown does not look very much unlike that of an incisor; its labial or external face is broader above than below,[1] and convex in both a transverse and a longitudinal direction, as in the incisors; the palatal or internal surface is also bevelled, and the lateral surfaces (mesial and distal), or those which lie adjacent to the contiguous teeth, are likewise somewhat triangular in form, but more rounded. In the superior canines a slight ridge descends upon the external or labial face from the summit of the terminal cusp to the neck, but is absent in the corresponding teeth below.

The internal or palatine aspect is slightly convex from side to side, but concave from above downward. The palatine convexity is occasioned by a well-marked vertical ridge which extends from the summit of the terminal cusp to the cingulum below; this latter structure is usually well defined, being stronger in the upper than in the lower teeth. There is, as a general rule, a prominent basal cusp at the junction of the two lateral ridges which connects with the vertical ridge, leaving a deep pit upon either side—a spot where caries very frequently occurs.

As already stated, the extremity of the crown slopes away upon either

[1] When the terms *above, below, superior,* and *inferior* are used in connection with a single tooth, they refer to the free as opposed to the implanted extremities: in the upper jaw the part of the crown which is really above is that which joins the root, but in the lower jaw it is the reverse of this. It is convenient to use these terms for all teeth, as they correctly apply to the lower teeth, so that when we speak of the superior extremity of the crown, the free or terminal part is meant, whether it belong to the upper or lower jaw.

side of the median cusp; that side which lies next to the premolars or bicuspids is longer than that which is directed toward the incisors, so that the distal or posterior moiety is greater than the mesial or anterior. This inequality of the two sides exists in both pairs of the canines, being less marked in the lower than in the upper; it furnishes a very useful rule by which a canine can be referred without difficulty to its proper side of the mouth.

The inferior cuspids or canines differ principally from those above in the shorter root, blunter median cusp, and less-marked posterior or lingual cingulum and basal cusp. The roots of both are thicker labially than lingually, and are generally traversed by a vertical groove upon either side.

The bicuspids or premolars are four in number in each jaw, and afford a further complication of the pattern of the crown by reason of the elevation of the basal cingulum into a strong internal cusp. In proportion as this part of the crown is well marked and complicated, there is a corresponding disposition to increase in the number of roots or fangs. These teeth, as their name implies, are provided with two cusps to the crown; those of the superior set are of subequal dimensions and considerably exceed the lower ones in size. The crown of the bicuspid, when viewed vertically, presents an imperfectly quadrate outline, which is most distinct in the second, and are broader than long. Two strong cusps, of which one is external and the other internal, occupy the grinding face, and are separated by a deep notch or valley, deepest in the centre. The anterior and posterior margins of this valley are bordered by slight ridges which connect the anterior and posterior extremities of the cusps; the anterior of these is a little more elevated than the posterior, and forms a useful guide in determining the mesial and distal surfaces of the tooth, and consequently the side of the jaw to which it belongs. In some instances the enamel forming the floor of the valley and adjacent sides of the cusps and ridges is quite smooth, but most frequently it is considerably wrinkled and thrown into a number of minor cusps and ridges, with intermediate indentations which offer receptacles for the lodgment of food.

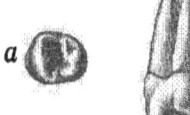

Fig. 220.

First Upper Bicuspid or Premolar of Man: *a*, vertical view of the crown; *b*, lateral view.

Of the two cusps, the external is slightly the larger and more elevated; it likewise has a greater antero-posterior extent. Its form is very much like the entire crown of the cuspid, terminating superiorly in a median cusp, from which the cutting edge gradually slopes away upon either side. The internal vertical rib is also present, but the external is absent. The internal or palatine cusp is thicker transversely than the buccal, and is more rounded. On account of the connecting ridges it has somewhat of a crescentic pattern.

Commonly, there is, to all appearance, but a single root, which is traversed upon the mesial and distal faces by vertical grooves which may unite near the apex, causing it to become divided. These vertical grooves are the external indication of two pulp-cavities in the implanted extremity, which unite about midway of the root, and are thence continued upward into the crown as a common cavity. The cavity thus formed is of greater transverse than antero-posterior[1] extent; in the vicinity of the neck it is little more than a narrow transverse fissure, which widens somewhat above, and is prolonged into two cornua corresponding to the two cusps. The external of these is the larger and most elevated.

While this condition of the roots and pulp-cavity is the one usually to be met with, nevertheless two roots are frequently found in the first bicuspid, and three roots are occasionally developed, two of which support the outer cusp; the pulp-cavity has then, of course, three divisions.

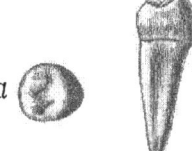

Fig. 221.

Second Lower Human Bicuspid: *a, b*, vertical and lateral views.

The principal differences between the upper and lower bicuspids or premolars are seen in the size of the internal cusp as compared with the external, the more cylindrical form of the root, and the almost complete absence of the vertical grooves, on account of which the pulp-cavity is, as a general rule, single. The crown consists of a large, somewhat conical external cusp, very convex without, to which is added a low lunate internal cingular ridge. The internal vertical ridge of the external cusp joins this cingulum near its central portion, leaving a deep pit upon either side where the destructive agencies of decay on the crowns of these teeth exhibit themselves most frequently. The degree to which this vertical rib is developed is subjected to great variation; it may be almost entirely absent in some individuals or strongly developed in others. The crown of the second or posterior bicuspid or premolar is more quadrate in outline than the anterior or first; the internal cusp is better developed, and frequently shows a tendency to form two.

The normal number of true molars is twelve, three on either side in each jaw, but, as already remarked, the last in both series may be absent. In a series of adult skulls of various civilized races which I have examined, twelve out of forty had one or both of these teeth wanting from the upper series, and in the lower jaw the proportion was ten to thirty. It is highly probable that in many of these cases these teeth had been present, but had disappeared early in life. Many examples could be cited in which the last or third molars wholly fail to be erupted, and it is established upon good authority that in many families one, two, or all of these teeth are habitually absent from generation to generation.

Fig. 222.

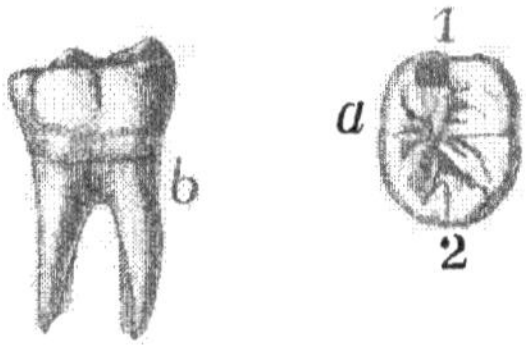

First Lower Human Molar: *a*, vertical view of the crown; 1, anterior; 2, posterior aspect; *b*, side view.

In the lower jaw the three molars in the more typical lower races are equal in size and substantially alike in pattern; their crowns are quadrangular in section, with the angles consider-

---

[1] By *antero-posterior* in this connection is meant the diameter which corresponds with the long axis of the jaw.

ably rounded off. They support four principal cusps, as in the quadritubercular molar generally, together with a fifth one behind, which is strictly homologous with the heel of these teeth in the more generalized members of the Primate section. These are separated by four distinct fissures arranged in the form of a cross; where the two limbs cross each other they widen out into a median valley deepest in the centre. The longitudinal of these, or the one which separates the external from the internal principal cusps, terminates in a posterior bifurcation which constricts off the fifth cusp or heel. The enamel lining this valley is, in perfectly unworn teeth, much corrugated, so that it is sometimes difficult to distinguish the principal cusps.

They are, with the exception of the last or third molar, implanted by two antero-posteriorly flattened roots, which join the crown at the moderately well-defined neck. These may be connate, having the two roots indicated only by a vertical groove upon either side. Each root is hollowed out in the centre to receive the radicular portion of the pulp, the cavity corresponding with the external form of the root. These unite into a common cavity above, which at about the time the tooth is erupted is relatively very large, but which becomes smaller with age, and is finally in old age obliterated through progressive calcification. The body of the cavity is terminated superiorly by cornua corresponding to the five cusps of the crown; of these the two anterior are most prolonged, and reach slightly above the inferior limit of the outer enamel covering.

In the higher races the last or third molar is usually smaller than the first and second, and does not have the cusp so well defined; but in many of the negro skulls I have examined it is nearly as large, and quite as well formed, as the two anterior to it. This tooth is more constant, both as regards presence and form, than the corresponding tooth above. It is, as a rule, two-rooted, but these roots may be confluent, in which case two vertical grooves mark a tendency in this direction.

The superior molars, like those in the lower jaw, are three in number, and have quadrituberculuar crowns normally, but many examples can be found in which the postero-internal cusp, the last one added in the quadrituberculuar molar, is little more than a cingulum,[1] and is scarcely entitled to the appellation of a cusp. In such cases it frequently has a position internal to the antero-internal cusp, and all stages between that and its normal position are to be met with.

The grinding face of the crown is of a squarish form, bearing a

---

[1] It is probable that this condition, of which I have seen a number of examples in the higher races, is a degenerate one, and is an effort to return to the tritubercular stage. Dr. Harrison Allen, in a communication to the Academy of Natural Sciences of Philadelphia, has recently called attention to the fact that in senile changes those structures which have been added last in the course of evolutionary growth are the first to disappear. Although this condition cannot be said to be in any way dependent upon individual senility, it is in all probability the result of senility of the race, wherein retrogressive modifications of any set of organs are first apparent in those parts which were the last to appear. It should be stated here that to Dr. Allen is due the credit of having prepared the way for all the more important generalizations that have been made in regard to the evolution of the quadritubercular tooth from the more primitive types. He demonstrated that the postero-internal cusp of the human molar is an outgrowth from the cingulum.

cusp at each angle. Of the two external, the anterior is slightly the larger, and is usually connected with the antero-internal by a strong ridge which skirts the anterior margins of the crown. The posterior is separated from it by a fissure which terminates internally in the median valley; it is also connected with the antero-internal cusp by a ridge, the oblique ridge. From its posterior margin a well-developed cingulum passes inward on the posterior border of the crown to join the postero-internal cusp, of which, as already remarked, this latter is a part.

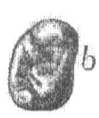

Fig. 223.

First Superior Human Molar: *a*, lateral, and *b*, vertical view.

The antero-internal cusp is the largest of the four, and by reason of its union with the cross-ridges above mentioned has a somewhat crescentic appearance. It is placed at the apex of a V which opens externally and encloses the median valley. The postero-internal cusp in the specimen figured stands a little posterior and internal to the last mentioned, being separated from it by a deep groove; it is little more than an enlargement of the strong posterior cingulum.

The roots are three in number, of which two are external or buccal, and support the two outer cusps, and one internal or lingual, supporting the two internal cusps. The two outer are not unfrequently connate, in which case the line of separation of the radicular portions of the pulp-cavities is indicated by a vertical groove. The palatine is the largest and longest root of the three.

While the structure here described usually obtains in the first and second molars, the last is more simple and variable. In the more civilized races it is exceptional for these teeth to be regular either in form or position, so great is their variability. The crown resembles in a general way those of the first and second molars, except that the oblique ridge is generally absent and the two internal cusps are blended together. The roots are connate and somewhat curved at their implanted extremity, and the pulp-cavity is single.

Occlusion of the Teeth.—The diagram (Fig. 224) on p. 446 represents the occlusion of the teeth. It has been previously stated that in a well-formed denture no one tooth rises higher than its fellows; that is, if the crowns of the teeth in position be turned, cusps and cutting edges, upon a plain or even surface, each tooth rests upon this surface. From this arrangement there is nothing to interfere with a perfect occlusion. Still, the fact must be recognized that while the above-described arrangement is true of a perfectly-developed jaw and teeth, yet so rarely is it found that it may be considered an *ideal denture.*

It has also been stated that the superior arch or row of teeth describes the segment of a larger circle than does the inferior row; this being the case, when the two are brought in contact, as in normally closing the mouth, the anterior superior teeth are thrown slightly over and anterior to the corresponding inferior teeth. Also with the bicuspids and molars, the external cusps of the superior ones are in closing slightly external to the corresponding cusps of the inferior. Another serviceable peculiarity is the noticeable absence of an exact opposition of tooth to tooth in closing, as will be seen by the diagram: the greater width of the superior central covers the width of the inferior central and a small portion of the inferior lateral; this brings the superior lateral over the remainder of the inferior one and the mesial fourth of the inferior cuspid, while the cusp of the superior cuspid fits into the concave space between the cusp of the inferior cuspid and the first bicuspid. In like manner, this

FIG. 224.

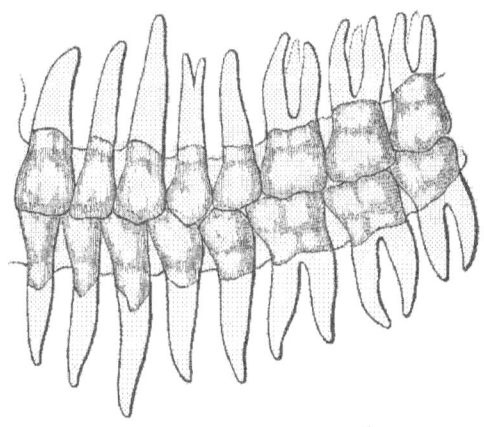

irregularity of opposition is maintained in all the teeth, so as to give each tooth a bearing on two teeth, except the superior third molar, which has but the corresponding tooth in the lower jaw for an antagonizer. This irregularity of opposition contributes to the efficiency of the teeth in mastication, and is a valuable feature when a tooth is lost from the arch in either jaw, for by this arrangement the tooth in partial antagonism with the one lost still maintains a portion of its usefulness by its occlusion with yet another tooth.

The *Deciduous or Temporary Teeth* (Fig. 225), twenty in number, are

FIG. 225.

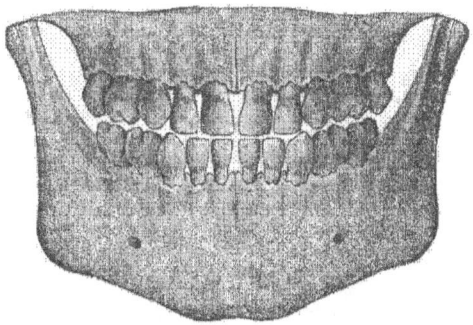

smaller than the permanent set, though they resemble them in their general conformation of crown and root, the bicuspids of the permanent set not being represented in the deciduous dentine. The formula, when they are normally developed, is I. $\frac{2}{2}$, C. $\frac{1}{1}$, M. $\frac{2}{2}=\frac{10}{10}=20$, the premolars or bicuspids being confined to the permanent set. A marked point of dissimilarity, as compared with their successors, is in the termination of the enamel on the neck of the tooth. In the permanent teeth the gradual completion of the enamel on the border of the cement marks but indistinctly the point of union of these two structures, while at the base of the deciduous crown the terminating enamel on the buccal and labial surfaces is recognized by a ridge or well-defined border which unmistakably marks its limitation and develops a well-constricted neck. This difference is often of importance in deciding as to whether a tooth in question belongs to the deciduous or permanent series. These teeth, from the fact that their crowns are largely calcified before birth, are much less liable to vices in conformation than their successors, but from deficient nutrition, want of use, and neglect not unfrequently become an easy prey to the ravages of dental caries. In common with a large class of the order to which man is closely allied, the difference in number between the deciduous and permanent set is twelve, the additional teeth being without predecessors.

The accompanying figure (226) represents the denture of a child about

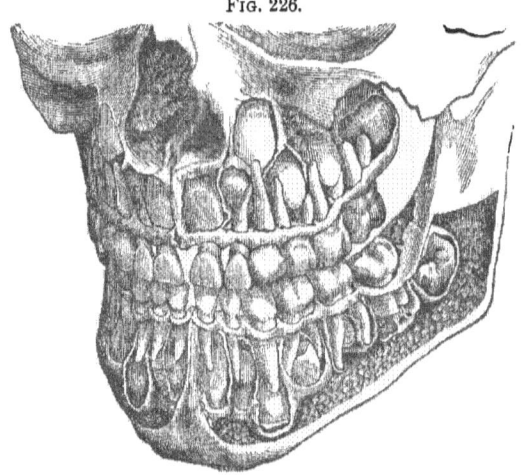

Fig. 226.

seven years of age. Twenty deciduous teeth, ten in each jaw, and the four first permanent molars, are erupted. The second permanent molar is seen in the crypt in the posterior part of each maxilla. Commencing with the median line, to the right of it and just above the erupted deciduous central incisor, we observe the permanent central with its crown fully calcified and the root partially so. In this case the crown of the permanent tooth stands in front of or on the labial side of the partially absorbed root of the deciduous central. This is not its constant relative position; not unfrequently the deciduous root is in front of the crown, as is seen in the adjoining lateral incisor. In this case the crown of the permanent lateral has the position which it invariably maintains. The crown of the permanent cuspid is here normally located quite above the root of its predecessor, and at the side of and in close proximity to the wing of the external nares. Next in position, a little below and slightly posterior to this cuspid crown, is that of the first bicuspid, this and its fellow, the second bicuspid, are located between the roots of their respective predecessors, the first and second deciduous molars. The same is true of those on the other side of the jaw, and, with slight variation, the same relative positions of the deciduous roots and permanent crowns are observed in the inferior maxilla. It is above stated that the figure represents the teeth of a child about seven years of age. The first permanent molars, it should be noted, are at this age erupted and in position, though their roots are not quite completed. The permanent central incisors will be the next to take their position at about the age of eight, followed by the laterals at nine, the first bicuspids at ten, the second bicuspids at eleven, the cuspids from twelve to thirteen, and the second molars from twelve to fourteen, which completes the eruption of the permanent teeth, with the exception of the third molars or wisdom teeth, these may take their position at eighteen or some years later. The anatomy of human dentition is further illustrated in plates placed at the end of this paper (see p. 505).

TEETH OF THE CARNIVORA.

Our knowledge of the philogenetic history of the unguiculate series has so increased within the last few years that it is now a matter of great difficulty to say just what forms should be included in the order *Carnivora*, as at present defined. If we take into account the living forms only, no one will hesitate in fixing its limit and giving to it a moderately good definition; but when the fossil representatives are considered, the interval between it and some of the contiguous orders, especially the *Insectivora*, is brought down to extremely small limits. We have already seen that the *Miacidæ* approach the dogs and civets in the *Carnivora* on the one hand, and the *Leptictidæ* of the *Insectivora* on the other. If a dog, bear, cat, and seal, all of which are admitted to belong to the *Carnivora*, be selected, and a careful comparison of their anatomical structure instituted, the differences between them will be found to be much greater than between such forms as *Stypolophus*, *Centetes*, *Miacis*, and the dogs and civets.

Every increment to our knowledge of the more exact relationship of the various groups seems to bring us nearer to the conclusion that our present classification is largely a matter of convenience, and often fails utterly to express the deeper and more important facts of origin and ancestry. Such reflections bring us abreast of the question, What is an order, a family, or a genus, etc.? And just here we approach a problem as to the solution of which no two naturalists agree.

It appears to me that the only way out of these difficulties is to consider the test of ancestry the only true basis of affinity. If it can be shown, for example, that any given assemblage of organic forms have descended from a common ancestor, however much they may differ among themselves, such a line or branch constitutes a natural division. Viewed from this standpoint, there can be little doubt that the order *Carnivora* represents the terminal extremities of several distinct branches, which arose not from one, but from two or perhaps three points in the *Insectivora*. The same reasoning holds good for many other orders.

The order *Carnivora*, as at present understood, is divisible into two sub-orders—*Fissipedia*, or the land carnivores, and the *Pinnipedia*, or aquatic flesh-eaters. The latter division includes the seals, sea-lions, and walruses, and is distinguished by the flipper-like modification of the feet for progression in the water, as well as by several important cranial characters. They are all known to be diphyodont, but the milk teeth disappear early; in some cases this occurs before birth, and in others a few weeks after. The teeth always possess comparatively simple crowns, which are either simple cones, as in the majority of the *Cetacea*, or laterally compressed, like the premolars of the dog, with smaller cusps along the edge, giving a well-defined serrated structure.

There are three families of this group, viz. the *Phocidæ* or seals, the *Otaridæ* or sea-lions and sea-bears, and the *Trichecidæ*, or walruses. In the common seal (*Phoca vitulina*), which is a good example of the first, the dental formula is I. $\frac{3}{2}$, C. $\frac{1}{1}$, Pm. $\frac{4}{4}$, M. $\frac{1}{1}$ = 34. The central pair of incisors above (Fig. 227) are the smallest, with sharp-pointed,

FIG. 227.

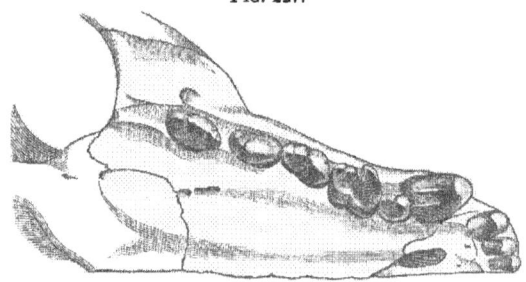

Vertical View of the Upper Jaw of a Harbor Seal (*Phoca vitulina*).

slightly hooked crowns; the next are similar in shape, but a little larger, while the outer pair are abruptly increased in size. These are separated from the canine by a diastema to admit the lower canine. The canine is a powerful tooth, with a conical recurved crown, and is deeply implanted in the maxillary bone. In the specimen figured, which is a young individual, it is remarkable for the very large size of the pulp-cavity, which extends nearly to the apex of the crown. The first premolar follows just inside and behind the canine, giving a crowded appearance to the first two premolars, the longitudinal axes of which are directed very obliquely to that of the succeeding teeth: it has no deciduous predecessor, as the corresponding tooth in the dog, and is one of the many examples in which it is difficult to say whether it should be relegated to the milk dentition as a persistent milk molar or whether it should be referred to the permanent set. It is implanted by a single root, also remarkable for the size of the pulp-cavity, and has a crown

Vol. I.—29

with a principal hook-shaped cusp, a small posterior basal cusp, and a strong internal cingulum.

The next three premolars are similar, except that they are larger, implanted by two roots, and have two posterior accessory cusps, the hindermost of which is very small. The single molar differs from the rest of the teeth in advance of it in having an anterior basal cusp, being relatively thicker at the base of the crown, and with a moderately well-defined internal cingulum, which displays a tendency to develop internal cusps.

The incisors and canines of the lower jaw are like those above, but the two incisors of each side are separated at the median line. The first premolar is single-rooted and somewhat larger than its fellow above. The crown displays a median cone with three posterior accessory cusps, and one very minute anterior one. The following teeth, including the molar, are all similarly constructed, but have the anterior basal cusp better defined.

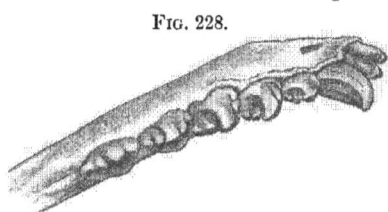

Vertical View of the Lower Jaw of a Harbor Seal.

Fig. 228.

In the hooded seals (*Cystophora*) the incisors are two upon each side above, and one upon each side below. The canines are comparatively large and powerful, while the molars and premolars are small and reduced to simple conical bodies, similar to the teeth of the cetaceans. In another genus (*Stenorhynchus*) the teeth are remarkable for the great length of the cusps, and in one species, *Leptonyx*, for the curvature of the accessory cusps toward the principal one, thereby resembling the trident of a fishing-spear.

A good example of the dentition of the *Otaridæ* is furnished by the fur seal (*Callorhynus ursinus*), which can usually be found in museums. The dental formula is I. $\frac{3}{2}$, C. $\frac{1}{1}$, Pm. $\frac{4}{4}$, M. $\frac{2}{1}$ = 36. The two median pairs of incisors above are subequal and laterally compressed. They each present a deep transverse notch in the summit of the crown, into which the incisiform extremities of the lower incisors bite; the outer pair are larger and sharp-pointed. The canines are relatively longer and more slender than in the seals, and have a well-defined posterior trenchant edge. The succeeding teeth are all alike in form and size, being implanted by single fangs. Their crowns are of a triangular shape when viewed from the side, and present a single cusp. It frequently happens that the bases of these teeth just where they emerge from the gums are very much eroded, the cause of which is not at present well understood.

Both the *Phocidæ* and *Otaridæ* are remarkable for their comparatively weak and slender jaws, the backward direction of the coronoid process, and the great distance intervening between its base and the last tooth. In the seals the palate is very broad posteriorly, and the last tooth does not extend behind the anterior root of the zygoma, whereas in the sea-lions the palate is long and narrow, and the last tooth is placed considerably behind the anterior termination of the zygomatic arch.

The *Trichechidæ* or walruses exhibit the most anomalous condition of the dental organs of any pinniped carnivore so far known, in that two enormous tusks are developed in the upper jaw, which occupy the position and fulfil the functions of canines. Owing to the transitory character of some of the other teeth, it is difficult to assign a definite dental formula to this animal. Prof. Flower makes it out to be I. $\frac{1}{2}$, C. $\frac{1}{1}$, Pm. $\frac{3}{3}$, M. $\frac{0}{0}$. Besides these there are, according to Tomes, several other small teeth to be found frequently in the position of the incisors, and he is disposed to regard them as the rudimentary representatives of the permanent normal ones in other animals; there can be little doubt that he is correct. Rudiments of true molars are also not unfrequently present in the back part of the jaws. The incisors and molars are small and simple, and are soon worn down even with the gums into obtuse oval grinding surfaces. The canines of the upper jaw protrude far below the level of the symphysis, and grow from persistent pulps. They are composed of dentine with a thin investment of cementum. Tomes says of them: "These great tusks are employed to tear up marine plants and turn over obstacles, the walrus feeding upon crustacea and also upon seaweed, etc.; they are also used to assist the animal in clambering over the ice; as they are of almost equal size in the female, they cannot be regarded as weapons of sexual offence, but they are undoubtedly used in the combats of the males."

The walruses and sea-lions agree with respect to the use of the hind limbs for progression on land, being able to walk on all fours fairly well; in the seals, on the other hand, the posterior members are rotated backward, and permanently fixed in this position, so as to be of little or no use in walking. In this respect they approach nearer to the cetacean condition.

Viewing the *Pinnipedia* as a whole, I am inclined to think that the relationship existing between them and the *Fissipedia* is more apparent than real; and although palæontology does not at present permit us to judge, I am of the opinion that they will ultimately be found to have been derived from an entirely different ancestry. Fossil remains are known as far back as the Miocene, but all that have so far been found are typically pinniped. The simple structure of the teeth finds a parallel in the *Insectivora* in the teeth of the lower jaw of the genus *Mesonyx*, already described, which Cope believes to have been more or less aquatic from the evidence afforded by some of the limb bones. This genus or an allied one may have been the progenitor of the pinnipeds, but too little is known of the skull-structure to say anything about the affinities between them.

The fissiped *Carnivora* are more extensive, both in number and variety, than the pinnipeds, and enjoy a wider range of distribution. Some of them are almost exclusively aquatic in habit, while others are arboreal, fossorial, or terrestrial. It is in this group that we meet with the highest specialization of the dental organs for the purpose of seizing, lacerating, and devouring living prey. In many the claws are extremely sharp and hook-shaped, and are provided with a special apparatus by which they are made retractile, thereby rendering them efficient organs of destruction and prehension as well. The feet are not modified into flippers, as in the pinnipeds, but constitute distinct "paws," which in the aquatic forms have webbed toes. The canines are always present and generally of formidable proportions, while the sectorial or shearing apparatus is present only in those that subsist exclusively on an animal diet.

They have been divided by Prof. Flower into three groups, which he has called the *Cynoidea*, *Ailuroidea*, and *Arctoidea*, defining them by the characters of the otic bullæ and the base of the skull. The first of these includes the dogs, wolves, and jackals, etc., and in all probability represents the central group. From it the civets, cats, etc., constituting the *Ailuroidea*, branch off on the one hand, while the bears, weasels, raccoons, etc. are closely connected on the other. The *Cynoidea* comprises two families—according to most authors only one; these are the *Canidæ*, or dogs, wolves, foxes, etc., and the *Megalotidæ*, including the single genus *Megalotis*, or the fennec of Africa, which, for reasons which will appear hereafter, I am strongly disposed to regard as an entirely distinct family.

A typical dentition of the *Canidæ* has already been described in that of the dog. About the only dental variations of importance to be seen in this family consists in the reduction of the number of premolars, addition or subtraction to the number of upper true molars to or from that of the dog, subtraction from the lower molar series, and slight modification in form of the sectorials. Upon these variations principally some thirteen or fifteen genera have been defined. The dental formula for many of the genera is the same as that of the dog, I. $\frac{3}{3}$, C. $\frac{1}{1}$, Pm. $\frac{4}{4}$, M. $\frac{2}{3}$ = 42, but the extinct Miocene genus *Amphicyon* (Fig. 229), found both in this country and Europe, had three true molars in the upper jaw. In another extinct genus (*Enhydrocyon*), described by Cope from the Miocene of the John Day beds of Oregon, the premolars are reduced to three in each jaw. *Oligobunus* is the name given by this author to another extinct genus from the same locality, in which the molar formula is Pm. $\frac{4}{4}$, M. $\frac{1}{2}$. The principal part of the skull is represented in Fig. 230. Still another genus of this family has been described by the same author from the rich fossiliferous deposits of that region under the name of *Hyænocyon*, which has three premolars above and below, with only a single true molar above.

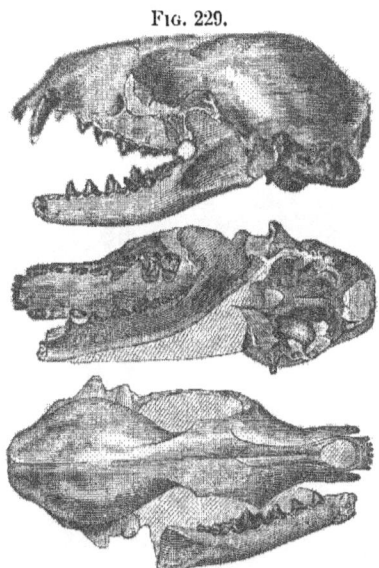

Fig. 229.

Skull of *Amphicyon cuspigerus*, Cope, with last superior molar lost, one-half natural size, from the John Day beds of Oregon (after Cope).

The sectorials of the more typical *Canidæ* are like those described in the dog, but in some genera—notably *Temnocyon* of Cope—the heel of the lower sectorial, instead of being basin-shaped, retains the more primitive structure, and consists of a single trenchant cusp (see Fig. 231).

In the extinct genus *Ailurodon* of Leidy the dental formula is the same as in the dog, but it approaches the cats, and especially the hyænas,

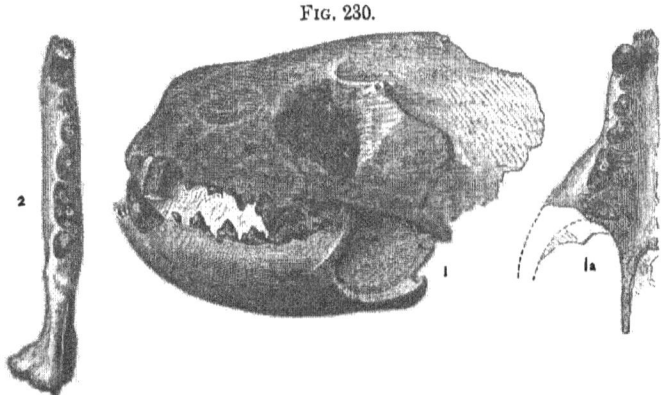

FIG. 230.

Portion of Skull of *Oligobunus crassivultus*, Cope, one-half natural size: 1a, right maxillary bone from below; 2, right mandibular ramus from above (after Cope).

in having three cusps to the blade of the superior sectorial, whereas the dog has only two. The premolars too are more robust than in the dog, constituting another approach to the condition of the *Hyænidæ*. The skull is represented in Fig. 232. The genus *Ichtitherium* of Gaudry (Fig. 233), from the Miocene of Pikermi, Greece, is an allied genus, but the third molar of the lower jaw is absent, leaving a formula, I. $\frac{3}{3}$, C. $\frac{1}{1}$, Pm. $\frac{4}{4}$, M. $\frac{2}{2}$ = 40. In one species (*I. robustum*) the last superior molars have nearly the same proportions as in the dog, while in another (*I. hipparionum*) the last molar is considerably reduced in size. It will thus be seen that these two genera depart from the central or typical *Canidæ*, and establish close connections with the *Hyænidæ*, which are closely affiliated with the cats and belong to the *Ailuroidea*. Cope has suggested that *Ailurodon* is the ancestor of the hyænas; and there is undoubtedly much evidence to support this opinion.

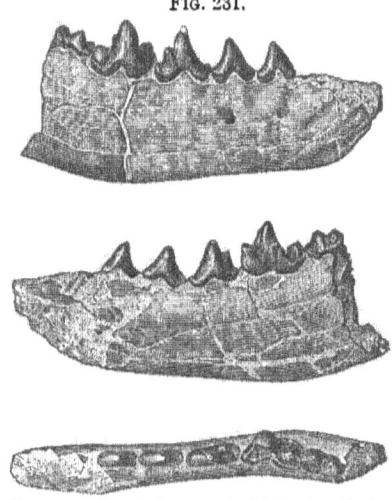

FIG. 231.

Temnocyon altigenis, Cope: part of Right Mandibular Ramus, one-half natural size, viewed from without, within, and above (after Cope).

The second family of the *Cynoidea* is the *Megalotidæ*, which is distinguished from the *Canidæ*—and, for that matter, from all other diphyodont monodelphous mammals—by the possession of *four* true molars in the lower jaw, thereby giving the formula I. $\frac{3}{3}$, C. $\frac{1}{1}$, Pm. $\frac{4}{4}$, M. $\frac{3}{4}$ = 46. The only other cases in which there are more than three true molars normally are found in the marsupials, edentates, and cetaceans; and in these two latter orders we have already seen that the teeth are not generally divisible into incisors, canines, premolars, and molars, on account of the development of only a single set. In the marsupials, however, as we shall presently see, the normal number of

Fig. 232.

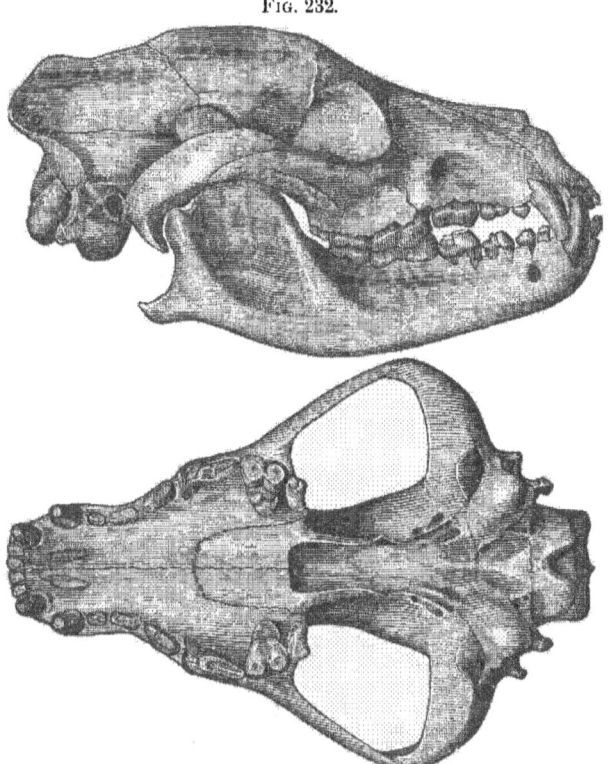

Skull of *Ailurodon sævus*, Leidy, three-eighths natural size (after Cope).

true molars is four, just as the number three is most common to diphyodont monodelphs. Reduction of the normal number is to be frequently observed in the monodelphs, and, as we have just seen in the *Canidæ*, occurs in genera otherwise nearly related; it cannot therefore be regarded as of more than generic importance, but there are no cases known to me in which teeth have been added. On the contrary, I am firmly of the opinion that not so much as a single tooth has ever been added to the diphyodont mammalian dentition in the course of development, but that specialization has invariably gone in the opposite direction, as almost all evidence of palæontology goes to show. The teeth are not otherwise remarkable, resembling distantly those of the dog in general pattern. The sectorials are not well defined, and the crowns generally have a tendency to the tubercular structure.

The second division of the Fissipedia (*Ailuroidea*) includes five families, the exact definitions of which the increasing knowledge of the extinct forms is tending every day to break down into hopeless confusion. The definitions are already very unsatisfactory and in many cases fail to define.

The families which approach nearest to the *Canidæ* are the *Hyænidæ* or hyænas, and the *Viverridæ* or civets. The evidence already cited brings the former of these families into the closest relationship with the central cynoid group. The dental formula of the existing hyænas (Fig. 234) is, I. $\frac{3}{3}$, C. $\frac{1}{1}$, Pm. $\frac{4}{3}$, M. $\frac{1}{1} = 34$. The incisors and canines have very much the same pattern as the corresponding teeth in the dog, as do also the premolars, with the exception of their more robust proportions

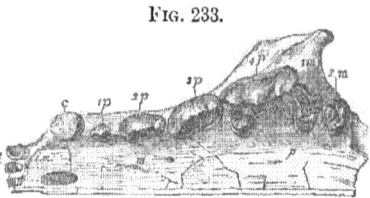

Fig. 233.

Superior Dental Series of *Ictitherium robustum*, two-thirds natural size (from Cope, after Gaudry).

Fig. 234.

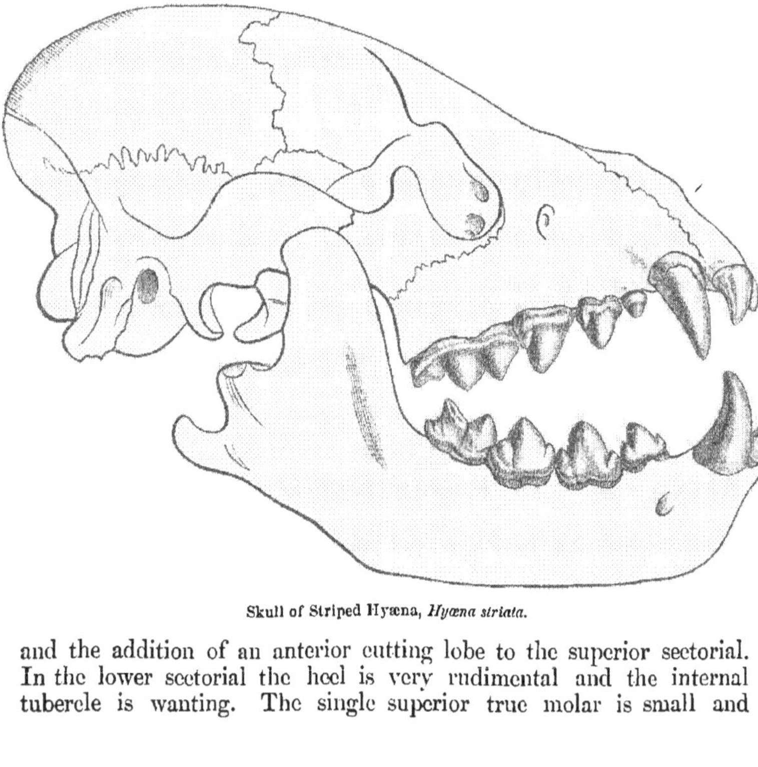

Skull of Striped Hyæna, *Hyæna striata*.

and the addition of an anterior cutting lobe to the superior sectorial. In the lower sectorial the heel is very rudimental and the internal tubercle is wanting. The single superior true molar is small and situated just internal to the posterior part of the great superior sectorial, so as to be completely hidden in an external view of the jaw.

In an extinct species (*Hyæna eximia*) there were four premolars in the lower jaw, giving the formula I. $\frac{3}{3}$, C. $\frac{1}{1}$, Pm. $\frac{4}{4}$, M. $\frac{1}{1}$ = 36. The inferior sectorial also has a well-defined heel. In the more ancient or Miocene representative of this family (*Hyænictis græca*, Fig. 235) the superior molar is much larger and has a more posterior position; the inferior sectorial (Fig. 236) has a relatively large basin-shaped heel, and there is a small second true molar behind it. It is through this genus that the transition is effected from the *Hyænidæ* to the *Canidæ* by way of *Ictitherium* and *Ailurodon*.

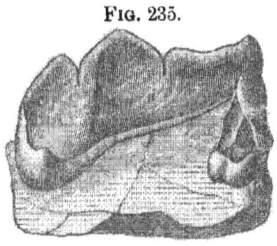

Fig. 235.
Superior Sectorial and First Molar of *Hyænictis græca* (after Gaudry).

The dental formula of the *Viverridæ* varies somewhat by reason of decrease in number of the premolars and molars. The more important of these will be noticed after we have first described the dentition of a typical example of the family, which is found in the genus *Herpestes*, or the mongoose. The dental formula is I. $\frac{3}{3}$, C. $\frac{1}{1}$, Pm. $\frac{4}{4}$, M. $\frac{2}{2}$ = 40. The incisors of the upper series have flattened oval crowns without lateral lobes, increasing in size from first to third; the canines are long, pointed, and recurved; the first three premolars have the usual pattern, but are devoid of accessory cusps. In the fourth premolar or superior sectorial the blade is composed of the usual two posterior cusps, separated by a fissure remarkable for its depth. There is also a rudimental anterior basal lobe, which arises from the cingulum. The internal lobe is unusually strong, and sends a trenchant ridge backward and outward to join the principal cone. The next tooth, or first true molar, is tritubercular, with two external and one internal cusp; the crown is remarkable for its transverse extent. The last molar is relatively small, and has a more internal position, possessing a bicuspid crown. The decrease in size of the true molars from that of the great sectorial, and the strongly inward curvature of the tooth-line behind, are more pronounced than in the dog, and altogether intermediate between that of the latter animal and the cats.

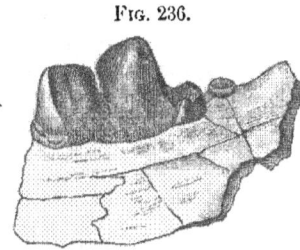

Fig. 236.
Fragment of Lower Jaw of *H. græca*, showing sectorial and second molar (after Gaudry).

The incisors of the lower jaw are smaller than the corresponding teeth above, and the summits of their crowns are distinctly notched; the canines are like those of the upper jaw, while the premolars have basal cusps which are largest behind. The first true molar or inferior sectorial furnishes a pattern intermediate between the tuberculo-sectorial and the well-defined sectorial. The primitive cone and anterior basal lobes are connected into a blade, the internal tubercle being large and furnishing the characteristic triangular appearance of this portion of the crown. The heel consists of a raised margin bearing several small tubercles. The last molar is quadritubercular, and seems to have retained the anterior triangle of the preceding tooth, together with one cusp of the heel. If this be so, it is an exception to the general rule, according to which the anterior cusp becomes obsolete.

In the two-spotted paradoxure (*Nandinia*) of West Africa the molar series is frequently reduced to M. ½, while in the bintourong (*Arctictis*) the last molar above and the first premolar below are often absent. The premolar formula of the genus *Galidea* is normally ¾, which likewise obtains in the kusimanse (*Crossarchus*) from the West Coast of Africa. The form of the inferior sectorial of the genus *Cynogale*, a Bornean representative of this family, is nearer that of a tubercular than a sectorial tooth. The three anterior cusps which go to make up the triangular portion are very much reduced, and have altogether lost their sectorial character; the superior sectorial, however, is much better defined as such. In another genus (*Eupleres*) the teeth are very small and the incisors stand far apart, on account of which, together with several cranial peculiarities, Dr. Gill gives it a distinct family rank.

It will thus be seen in a survey of the dental organs of this family that they are almost identical with the genus *Didymictis* of our American Eocene, which has already been described, and I think there can be little doubt that they are the derivatives of this or some nearly related genus.

Another family, which stands intermediate between the civets and cats, is represented by the single living genus *Cryptoprocta*, which is limited in its distribution to the island of Madagascar. Some authors classify it as a sub-family of the cats, others as a sub-family of the civets, while others again make it a distinct family. No better argument, it seems to me, could be advanced in support of its intermediate nature. It undoubtedly has strong affinities with both families, and goes far toward bridging over the interval between them. The recent discoveries of Cope and Filhol have shown it to be the surviving remnant of an extensive group which lived in this country and Europe, and which were the ancestors of the cats, and in all probability the derivatives of the more generalized civets. In distinguishing between the *Felidæ*, *Viverridæ*, and *Cryptoproctidæ* the foramina at the base of the cranium afford the best, if not the only, grounds for separation. Previous to our knowledge of the extinct forms the number of the molar teeth was also used for this purpose, but owing to the intermediate condition of this latter character in many of the fossils it must be abandoned as altogether worthless. In the *Cryptoproctidæ* the alisphenoid bone is perforated by a canal—the alisphenoid canal—for the passage of the external carotid artery in its course forward. The foramen for the entrance of the internal carotid in its passage to the brain is also well defined, and of a considerable size. In the *Felidæ* there is no alisphenoid canal, and the carotid canal is minute or absent. In the *Viverridæ* the alisphenoid canal is generally present, but not invariably so; the foramen for the entrance of the internal carotid is of moderate proportions, as in the *Cryptoproctidæ*, from which I can see no very good reasons for distinguishing them as a family. Cope associates a number of extinct genera together under the name of *Nimravidæ*, and defines them from the *Felidæ* by a number of characters in which they agree with the *Cryptoproctidæ*; the distinctions between them and this latter family are not so apparent. The dental formula of *Cryptoprocta* is I. $\frac{3}{3}$, C. $\frac{1}{1}$, Pm. $\frac{4}{3}$, M. $\frac{1}{1} = 34$. The incisors and canines resemble those of the cats generally; the first premolar in the upper jaw is caducous, and does not usually appear in the adult skull. The superior sectorial has a rudimental anterior basal lobe, an internal tubercle, and a well-defined blade. The molar is a much smaller tooth, and has an internal position, as in the hyænas. In the lower jaw the sectorial has a faint heel and lacks the internal tubercle, and is altogether feline in its appearance.

The following extinct genera are enumerated and defined by Cope as belonging to the family *Nimravidæ*:[1]

I. Lateral and anterior faces of mandible continuous; no inferior flange.
  a. No anterior lobe of superior sectorial; inferior sectorial with a heel; canines smooth.
  Pm. $\frac{4}{4}$, M. $\frac{1}{1}$; inferior sectorial with internal tubercle . . . . *Prœlurus*.
  Pm. $\frac{3}{3}$, M. $\frac{1}{1}$; inferior sectorial without internal tubercle . . . . *Pseudælurus*.
  Pm. $\frac{3}{3}$, M. $\frac{1}{1}$; inferior sectorial without internal tubercle . . . *Cryptoprocta*.[2]
II. Lateral and anterior faces of mandible separated by a vertical angle; no inferior flange; incisors obspatulate.
  a. No anterior lobe of superior sectorial; inferior sectorial with a heel (and no internal tubercle); incisiors truncate.
  Pm. $\frac{4}{4}$, M. $\frac{1}{1}$; canine smooth . . . . *Archælurus*.
  Pm. $\frac{3}{3}$, M. $\frac{1}{1}$; canines denticulate . . . . *Ælurogale*.
  Pm. $\frac{3}{3}$, M. $\frac{1}{1}$; canines denticulate . . . . *Nimravus*.
III. Lateral and anterior faces of mandible separated by vertical angle; an inferior flange; canines denticulate.
  a. No or a small anterior basal lobe of superior sectorial; inferior sectorial with a heel. No posterior lobes on crown of premolars.
  Pm. $\frac{3}{3}$, M. $\frac{1}{1}$ . . . . *Dinictis*.
  Pm. $\frac{3}{3}$, M. $\frac{1}{1}$ . . . . *Pogonodon*.
  Pm. $\frac{2 \text{ or } 3}{2}$ M. $\frac{1}{1}$ . . . . *Hoplophoneus*.
  Pm. $\frac{?}{1}$, M. $\frac{?}{1}$ . . . . *Eusmilus*.

Prolæurus is known to have possessed five digits in each foot, as *Cryptoprocta*, and it is probable that two sub-families should be made, since others had only four in the pes.

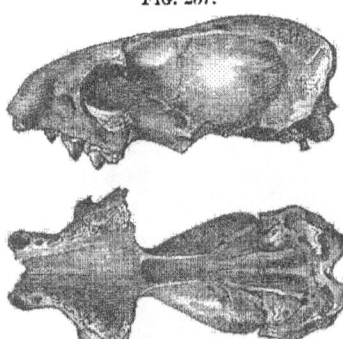

Fig. 237.

Prœlurus julieni, Filh., two-thirds natural size.

The dentition of *Prœlurus* (Fig. 237) is more primitive than *Cryptoprocta* in the following characters: there are four premolars in the lower jaw; the superior sectorial has no anterior basal lobe; the inferior sectorial has a strong heel and an internal tubercle; and there are *two* true molars below.

Pseudælurus agrees more nearly with *Cryptoprocta*, but lacks one premolar in the upper series. As already observed, the first premolar is caducous in this latter genus, and they may be the same. In the sec-

---

[1] "On the Extinct Cats of America," *American Naturalist*, Dec., 1880.
[2] I have combined the *Nimravidæ* and the *Cryptoproctidæ*, and have inserted this genus where it seems to most appropriately belong.

ond section the anterior and lateral faces of the mandible are separated by an angle or vertical ridge, which gives to the jaw the appearance of having a square chin.

The genera of this section, with the exception of *Ælurogale*, are from the Miocene beds of the John Day Valley, Oregon, and were described by Cope; the premolar series shows a gradual reduction in number, but they all retain the heel to the inferior sectorial and the generalized character of two true molars in the lower jaw. *Archælurus* and *Nimravus* are represented in the accompanying figures, 239 and 240.

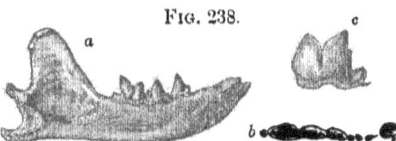

Fig. 238.

Prœlurus julieni, Filh., two-thirds natural size: *a*, inner view of mandible; *b*, superior view of inferior teeth; *c*, inferior sectorial, natural size (from Cope after Filhol).

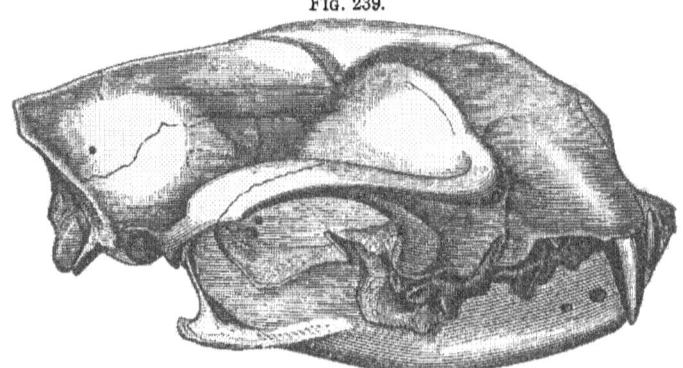

Archælurus debilis, Cope, Skull, one-half natural size (after Cope).

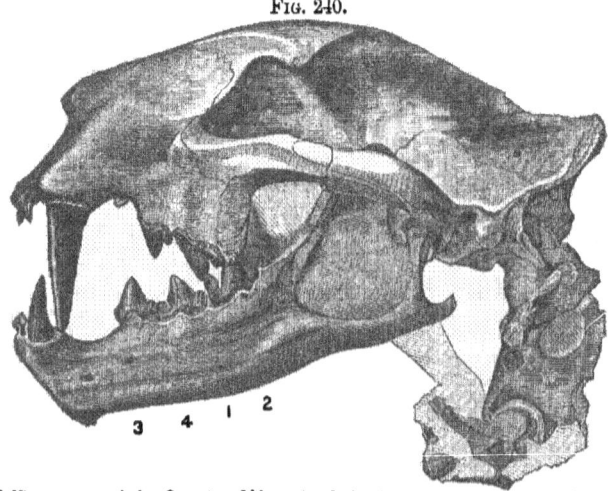

Fig. 240.

Skull of *Nimravus gomphodus*, Cope, two-fifths natural size (after Cope): 1, 2, first and second true molars; 3, 4, third and fourth premolars of lower jaw.

460   DENTAL ANATOMY.

In the third section the mandible possesses a strong inferior flange upon each side to protect the powerful canines of the upper jaw, which in some forms project far below the level of the symphysis. They are therefore known as the "sabre-tooth division." In the first of these genera, *Dinictis* (Fig. 241), the true molars are ½, the inferior sectorial

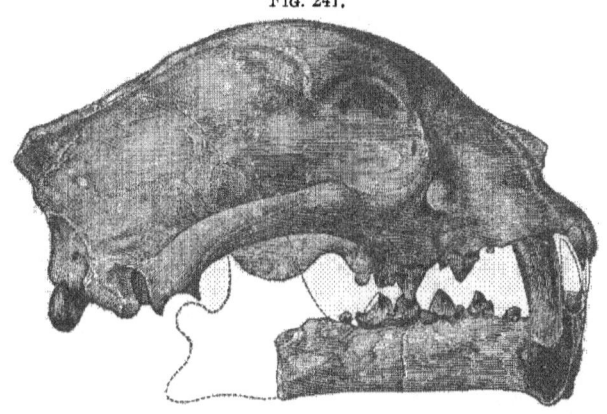

FIG. 241.

Skull of *Dincitis cyclops*, one-half natural size (after Cope).

has a heel, and the true molar above is a moderately well-developed tooth, as in the preceding genera. The genera *Hoplophoneus* and *Pogonodon* carry dental specialization several steps further, while in *Eusmilus* we have the highest point reached by any of this group, which is in many respects superior to the living cats.

Cope, in commenting upon the dentition of this group, says: "It is readily perceived that the genera above enumerated form an unusually simple series, representing stages in the following modifications of parts: (1) In the reduced number of molar teeth; (2) in the enlarged size of the superior canine teeth; (3) in the diminished size of the inferior canine teeth; (4) in the conic form of the crowns of the incisors; (5) in the addition of a cutting lobe to the anterior base of the superior sectorial tooth; (6) in the obliteration of the inner tubercle of the lower sectorial, and (7) in the extinction of the heel of the same; (8) in the development of an inferior flange at the latero-anterior angle of the front of the ramus of the lower jaw; (9) in the development of cutting lobes upon the posterior border of the large premolar teeth. . . . . The succession of the genera above pointed out coincides with the order of geologic time very nearly. . . . . The relations of these genera are very close, as they differ in many cases by the addition or subtraction of a single tooth from each dental series. These characters are not even always constant in the same species, so that the evidence of descent, so far as the genera are concerned, is conclusive. No fuller genealogical series exists than that which I have discovered among the extinct cats."

The last family of the *Ailuroidea* is the *Felidæ*, in which we meet with the highest point in specialization that has been reached in the flesh-eating Mammalia. It includes two divisions—one in which the superior canines are normal and without the vertical angles and inferior flanges to the mandible; and another, "sabre-tooth division," wherein the superior canines are enormously enlarged, denticulate, and protected by inferior flanges of the rami.

The first of these groups or sub-families is the more generalized, and embraces all the existing cats or those animals popularly known as lions,

Fig. 242.

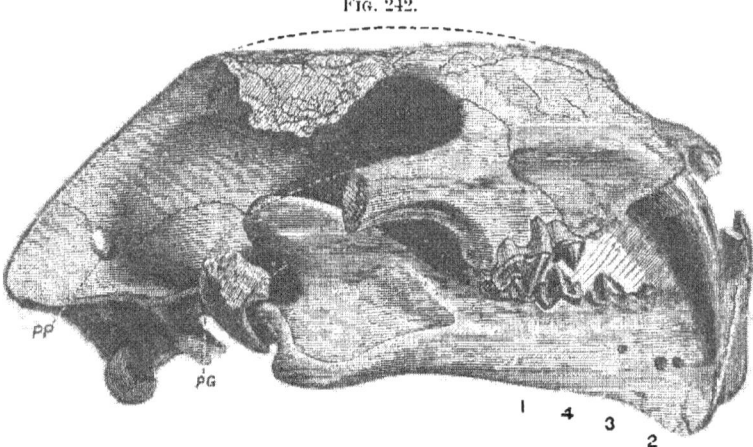

Pogonodon platycopis, Skull, less than two-fifths natural size (after Cope): 2, 3, 4, second, third, and fourth premolars, and 1, first molar of lower jaw; *P G*, post-glenoid foramen; *P P*, post-parietal foramen.

tigers, leopards, panthers, etc. Five genera have been established in this division on characters of the teeth and orbit. It is here that the domestic cat belongs, and its dentition may be taken as a good average representation of that of the sub-family.

The dental formula in this animal is I. $\frac{3}{3}$, C. $\frac{1}{1}$, Pm. $\frac{3}{2}$, M. $\frac{1}{1} = 30$. The incisors are relatively small, and are disposed almost transversely across the front of the jaw. The first premolar above is a small, single-rooted tooth, and is situated at a considerable distance from the canine, which has the usual form and proportions of that tooth in the *Carnivora* generally. The second is larger and two-rooted, while the fourth or upper sectorial is decidedly the largest tooth of the superior series; it has three external cusps united into a blade, and a small internal tubercle. The single molar is very small and functionless, being placed internal to the posterior part of the large sectorial. In the lower jaw the premolars are proportionately large, having two fangs and posterior accessory cusps. The sectorial is specialized, and consists simply of two cusps forming a trenchant blade; both the heel and internal tubercle are absent.

The lynxes have one less premolar upon each side above than the cat, and for this reason are placed in a distinct genus. In the flat-headed cat and the fishing cat the orbit is completely encircled by bone—an unusual occurrence in this family. In both, the number of teeth is the same as in the domestic cat, but in the former the first premolar in the upper jaw has a single fang, whereas in the latter this tooth is two-rooted. Upon these characters two genera have been established. The clouded tiger of India has a dental formula like that of the lynxes, and approaches the "sabre-tooth division" in the enlargement of the superior canines, by reason of which it has also been given a generic rank. The hunting leopard, or cheetah, forms another genus, and is distin-

Fig. 243.

Cranium of *Smilodon necator*, Gervais, one-third natural size (after Cope).

guished by the absence of the internal tubercle of the superior sectorial. All the other cats are very much alike, and can be distinguished from one another only specifically, being classified, therefore, under the genus Felis.

The second division is extinct, despite the fact that they reveal to us the most perfect laniary dental apparatus yet known within the limits of the Carnivora, and were of the most formidable size. Two genera are known, of which the cranium of one (*Smilodon*) is represented in Fig. 243. In this animal the dental formula is I. $\frac{3}{2}$, C. $\frac{1}{1}$, Pm. $\frac{2}{2 \text{ or } 1}$, M. $\frac{0}{1}$ = 24 or 26, and marks the extreme point in dental specialization in this order, as far as reduction is concerned. The canines of the upper jaw are of prodigious size in comparison with those of the lower series, having compressed crowns with serrulate edges. The superior molar has disappeared, and the first premolar in the lower jaw in some species is wanting. The exact use of the great superior canines is not very clearly understood. The possession of retractile claws, the reduction of the molar and premolar series, together with the general perfection of the sectorial apparatus, are strictly in keeping with a most carnivorous habit; but with all this it must have been impossible for the animal to open its mouth wide enough to take a firm grip upon a living prey, on account of the great length of the upper canines.

Seeing that in the existing cats their chief destructive powers reside in their biting qualifications, it is difficult to understand how these animals inflicted wounds sufficient to destroy their prey, unless they did so with that part of the tusk which projected below the level of the symphysis when the mouth was closed, just as the walrus uses his tusks to clamber over the ice. They may also have been used to assist the animal in climbing, and in this way attained their great size.

The animals composing the last group, *Arctoidea*, are the least carnivorous, and do not as a general rule display as trenchant and sectorial dental organs as the two preceding; in two families the almost exclusively carnivorous habits are manifested by sectorials of moderate perfection; this condition is associated with a reduction of molars and premolars from the number possessed by the dog. In the others the molars are more or less tubercular—a structure better fitted for the mastication of the mixed diet upon which they subsist—and usually exceed the premolars in size and strength. The extremes of dental variation in this group are exhibited by the bears and weasels, of which the former are the farthest and the latter the least removed from the more typical carnivores in the structure of the teeth.

In the bears the dental formula is the same as in the dog, but in most of the living species the three anterior premolars are very small, and frequently disappear in old age, leaving a wide space between the fourth and the canine. In the upper jaw the teeth progressively increase in size from the fourth premolar to the last molar, which, besides being quadritubercular, is provided with a large posterior heel rounded off behind; by the addition of this heel the crown is rendered elliptical in transverse section, the antero-posterior diameter being twice that of the transverse. The first true molar has four cusps on its triturating face, and is subquadrate in outline; the fourth premolar is tricuspid, as in the dog, but the two outer cusps are not united into a perfect blade, and the internal lobe is large and has a median position. This tooth is relatively small, and is situated considerably in advance of the canthus or angle of the mouth; it is doubtful whether its possessor ever makes use of it as a sectorial organ, but rather prefers to tear the tough animal membranes than to divide them with the sectorials, as the dogs and cats do.

In the lower jaw the first true molar betrays the same lack of carnivorous specialization as the upper teeth, being essentially tubercular in structure, although the proper elements of the sectorial of the dog can be easily made out; the crown is much elongated, and is narrower in front than behind, the heel composing at least half of the crown. The next tooth behind it is the largest of this series, and is perfectly quadritubercular; the last molar is smaller, with a subcircular grinding face, upon which the tubercles are poorly defined.

While the structure here described is found in all the northern more carnivorous bears, the tropical frugivorous species retain to a greater extent the integrity and more normal condition of the anterior premolars. This is especially apparent in a genus recently discovered in the mountains of oriental Thibet and described under the name of *Æluropus*. In this animal the first premolar only is small, while the others gradually increase in size to the last molar, which has a comparatively small heel.

The palæontological evidence is as yet too meagre to demonstrate with any considerable degree of certainty the evolution of this group of the *Carnivora*, but some suggestive hints of their former connection with the *Cynoidea* are afforded by the extinct genus *Hyænarctos*, which was originally described by Dr. Falconer from the Sewalik Hills in India. This genus displays three premolars of normal proportions and a large sectorial, together with a last superior molar in which the heel is absent. In one species in particular, *H. hemicyon*, from the Miocene of Sansan in France, which is provisionally referred to this genus, the two true molars in the upper jaw have about the same proportions as in the dog, and otherwise resemble them very much. The sectorial, as is indicated by the roots, was large, with the internal tubercle placed opposite the middle part of the crown. If it were not for this latter fact, the fragment of the upper jaw upon which the species was established would readily pass for that of a member of the *Canidæ*.

In the weasels, which constitute the family *Mustelidæ*, the sectorials are well defined as such, and some of them, notably the typical weasels, possess retractile claws. In none does the molar formula exceed $\frac{1}{2}$, except a fossil genus, *Lutrictis*, a near ally of the otters, in which the molars are two in the upper jaw; it may, however, be reduced to $\frac{1}{1}$, as in the case of the Cape ratel (*Mellivora capensis*). The premolars vary in number, as do also the sectorial in structure.

The dentition of the American pine marten (Fig. 244) will serve as an illustration of this family, although it is somewhat more specialized in a carnivorous direction than most of them. Its dental formula is I. $\frac{3}{3}$, C. $\frac{1}{1}$, Pm. $\frac{4}{4}$, M. $\frac{1}{2}$ = 38. The incisors, canines, and premolars have approximately the same structure as those of the dog, except that the fourth premolar or superior sectorial has the two outer cusps blended together, with the vertical notch absent. The true molar is tubercular, and has a greater transverse than longitudinal extent. In the lower jaw the sectorial is very much like that of the dog,

TEETH OF THE VERTEBRATA.

while the second molar is small, single-rooted, having a crown with one cusp.

The teeth of the raccoons and allied forms are intermediate between those of the bears and weasels in many respects, with a stronger tendency to the tubercular than to the sectorial pattern.

TEETH OF THE CHEIROPTERA.—The modification of the anterior members for flight distinguishes this order from all other unguiculates at once. Excluding this peculiarity, which is universal among them, they are closely related to the *Insectivora*, and without doubt have been derived from some arboreal representative of this order. It is conceivable that in jumping from branch to branch they have first developed a lateral fold of integument, similar to that seen in the flying squirrels, which later involved the fore limbs and extended to the neck. The flying lemur (*Galeopithecus*) furnishes such a transitional condition, both in the possession of the membrane and the elongated and slender fore limbs, although it is highly improbable that the bats have descended through this genus.

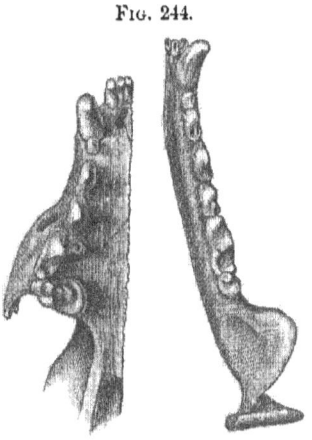

FIG. 244.

Vertical View of the Upper and Lower Jaw of American Pine Marten (*Mustela americana*).

The incisors are never more than two upon each side above, while the lower jaw is usually provided with the same number, but may be increased to three. Canines are always present in both jaws, but are of variable proportions. In the insect-eating forms (*Animalivora*), which includes the great bulk of the species, the upper molar teeth invariably display the peculiar W-pattern of the moles, shrews, etc. of the *Insectivora*, already noticed. The premaxillary bones are always small, and seldom meet in the median line so as to leave the tooth-border interrupted in front. In the W-pattern of the superior molars, the absence of the median pair of upper incisors, and the small premaxillaries it is interesting to note the resemblances they bear to the squirrel shrews (*Tuupaiadæ*), the only other insectivores besides the flying lemur which are known to be arboreal in habit. Ignorance of the rest of the anatomy of this genus does not permit me to state whether it strengthens this resemblance or otherwise, but upon the whole I am inclined to believe that some such arboreal insectivore was the ancestor of the bats.

The dentition of the blood-sucking vampires is modified in accordance with their habits, as is also the entire alimentary canal, and deviates quite extensively from the normal condition of that of the insect-eaters. The alimentary tract consists of little more than a straight tube from mouth to anus, and is thus adapted to the assimilation of the blood of living animals, upon which it feeds.

The large incisors of the upper jaw are two in number, one upon each side, whose roots extend into the maxillary bone, and whose compressed, sharp-pointed, hook-shaped crowns are specially fitted to puncture the skin of an animal sufficiently to cause the blood to flow freely. The canines are almost equal in size and similar in shape, while the lower incisors and canines are small. The molars are reduced to two in the upper and three upon each side in the lower jaw; the upper molars are implanted by single fangs and have simple conical crowns; in the lower jaw the first two are like those above, but the third has two fangs and a bilobed crown, and is considered by Owen to be homologous with the last premolars of insectivorous bats. The dental formula is thus reduced to I. $\frac{1}{2}$, C. $\frac{1}{1}$, Pm. $\frac{2}{3}$ = 20.

The frugivorous bats, which are popularly known as "flying foxes," offer another deviation from the usual structure in the pattern of the molar teeth; those in the upper jaw have crowns of a subcircular form in outline with a central longitudinal depression, upon each side of which the edge is elevated into a cusp. Those of the lower series are similar but smaller, with the cusps more pronounced and the median groove narrower.

TEETH OF THE RODENTIA.—The amount of minor variation in the dental organs of this order is so extensive that their complete elucidation is hardly within the scope of the present work; a description of the leading types must suffice. That which most conspicuously distinguishes the rodents from all other mammals is the possession of two powerful curved incisors in each jaw, which grow from persistent pulps and are faced with enamel. The roots are implanted deeply in the substance of the jaw bones; in the lower jaw, often reaching as far back as the coronoid process. In consequence of the distribution of the enamel upon the front face of the tooth, leaving the dentine naked behind, the inequality of wear between the two surfaces is always marked, and constantly preserves a chisel point to the crown—a structure pre-eminently adapted to the gnawing habits of its possessor. Concomitant with this modification the canines are always absent, the premaxillary bones are large to support the roots of the incisors, and there is a wide space between the first molar or premolar and the incisor, in which no teeth appear. The mandibular condyle, moreover, is globular in form and never transverse, thereby allowing excursion only in an antero-posterior direction.

The order thus distinguished is divisble into four sub-orders, of which the rat, squirrel, porcupine, and rabbit are typical representatives of each.

The dental formula of the common rat (Fig. 245) is I. $\frac{1}{1}$, C. $\frac{0}{0}$, Pm. $\frac{0}{0}$, M. $\frac{3}{3}$ = 16. Deciduous teeth are entirely wanting, and it is therefore monophyodont—a condition which we would be led to anticipate, as far as the molar and premolar series is concerned, in view of the subtraction of the latter. The absence of any deciduous predecessors of the two pairs of incisors is said to be a constant feature of all rodents except the hares, so that monophyodontism of this highly heterodont animal need not occasion surprise.

FIG. 245.

Cranium of Common Rat, *Mus decumanus.*

The incisors are of the usual pattern displayed by the order—large, curved, compressed teeth, with chisel-shaped crowns, which are stained a deep orange color on the anterior face; the pigment which produces this color is intimately incorporated with the enamel itself, as in the shrews, and serves to sharply define the limits of the enamel covering.

In both the upper and lower jaws the first molar is the largest and the third the smallest. They are implanted by distinct roots, the opposite rows of teeth being nearly parallel. The crowns are made up of three curved transverse ridges, with the convexity in front and the concavity behind; the two anterior of these, in the upper teeth, are terminated internally by well-marked cusps, which rise above the summits of the ridges. In the last tooth the anterior and posterior of these ridges are less distinctly marked, and are reduced to little more than internal tubercles. The second molar of the lower jaw has the last crest rudimental, and in the third it is entirely wanting.

While this structure prevails in the teeth of the more typical murines, others possess molars with crowns of much greater complexity and without roots. Such is exemplified by the arvicoline section of the *Muridæ*, in which the crown is cleft to the median line by vertical fissures upon each side placed alternately. The structure of the grinding surface which results from this arrangement is a system of alternate triangular prisms connected in the middle of the crown by a narrow band of dentine. This is well shown in the accompanying figure.

Fig. 246.

Vertical View of the Grinding Surface of the First Lower Molar of a Muskrat (*Fiber zibethicus*).

In the squirrel division premolars are always present, in consequence of which there are deciduous teeth. With the exception of the beaver family, the teeth are very similar in the different species, the only important variation occurring in the number of premolars. In the common fox squirrel the dental formula is I. $\frac{1}{1}$, C. $\frac{0}{0}$, Pm. $\frac{1}{1}$, M. $\frac{3}{3} = 20$. The incisors (Fig. 247) are not so robust as in the rat, and, like them, are colored upon the anterior face.

Fig. 247.

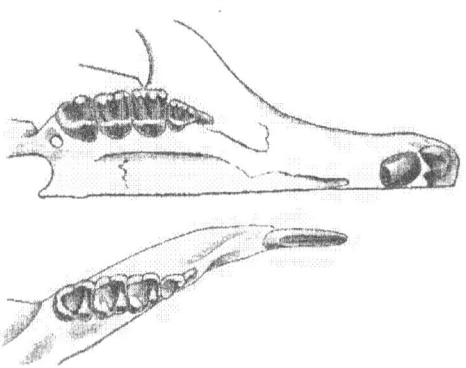

Vertical View of the Teeth of Fox Squirrel (*Sciurus carolinensis*).

The first and only premolar is smaller, implanted by three roots, and has a triangular tricuspid crown. The three true molars in the upper jaw are larger and subequal in size. Their crowns are imperfectly quadrate in outline, and provided with two transverse crests which join a large marginal cusp on the internal border. There is in addition an anterior and posterior cingulum, which becomes continuous with the large marginal cusp. The inferior molars have the same quadrate outline as those above, but the crowns present a central depression surrounded by a slightly elevated margin bearing a cusp at each angle.

In this sub-order is to be found the nearest approach to the quadritubercular condition of the molar teeth in any of the *Rodentia*, in consequence of which there is little difficulty in comprehending their organization; but when we come to analyze the highly complex form of molar which some of the porcupines exhibit, we naturally seek for a key to a solution of their structure on the basis of the quadrituberculars; this is all the more natural when we remember that the squirrels present the oldest known representatives of the order in the genus *Plesiarctomys* of Middle Eocene Age, which scarcely differs generically from the living forms. The teeth of the American porcupine (*Erethizon*), while possessing in general the molar pattern of the squirrel, nevertheless differs from it sufficiently in the direction of the more specialized hystricine teeth to let us into the secret of how these complex forms have arisen from that of the squirrel.

In the description of the molars of the squirrel we have already seen that the face of the crown is marked by three transverse ridges, enclosing two valleys, which open externally and are bounded internally by a thick marginal cusp. Now, in the first premolar of the porcupine (Fig. 248) which is unusually instructive, the three transverse ridges are present, but considerably augmented in height, together with a fourth ridge behind added from the cingulum. The valleys separating the three anterior crests open externally, while the fourth coalesces externally with the third, so as to enclose a deep pit or fossette; the strong internal marginal cusp is likewise present, but is interrupted on its inner side by a deep wide valley which opens internally. The succeeding molars are like it in structure, except that the first and second transverse ridges unite at their extremities to form a second fossette in front.

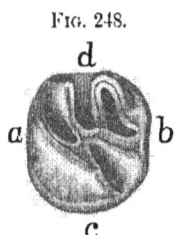

FIG. 248.

First Lower Premolar of Porcupine (*Erethizon dorsalis*), vertical view: *a*, anterior; *b*, posterior; *c*, internal; and *d*, external surfaces of the crown.

In other genera of this group the valleys are still further deepened by the elevation of the ridges, and other indentations are added from within. As a protection against fracture of the now laminar crests, cementum is added, which completely fills up the valleys, leaving the grinding surface approximately smooth. This is the condition attained by the beaver among the sciuromorph or squirrel sub-order, as well as a majority of the hystricomorphs or porcupines. As a further complication in this series, the external and internal valleys unite across the face of the crown, leaving transverse laminæ connected only at the base and bound together above by cementum; of which the guinea-pig is an example. Finally, the extreme of specialization is reached in the capybara, wherein these transverse laminæ are as many as thirteen or fourteen in a single tooth.

This point of perfection rivals that of the elephant, and is undoubtedly a long way removed from the quadritubercular structure. On account of this highly complex molar dentition and certain cranial peculiarities Dr. Gill has proposed to give this genus a distinct family rank.

The last sub-order of the *Rodentia* is the *Lagomorpha*, which includes the hares and rabbits. The dental formula in this group is constantly I. $\frac{2}{1}$, C. $\frac{0}{0}$, Pm. $\frac{3}{2}$, M. $\frac{3}{3} = 28$, in addition to which in very young specimens there is another or third pair of incisors in the upper jaw to be added to the permanent set. Huxley has recently shown that the deciduous dentition is D. I. $\frac{3}{2}$, D. M. $\frac{3}{2}$, which brings this group of the *Rodentia* into strict accord with other Mammalia in the replacement of the teeth.

FIG. 249.

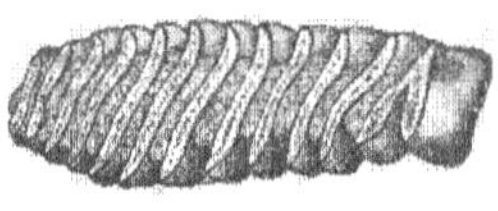

Last Molar of Capybara (*Hydrochærus capybara*), vertical view.

The median pair of superior incisors depart from the usual pattern, inasmuch as they are indented upon their anterior faces by a vertical groove near the middle of the tooth; they are otherwise as in the genera already noticed, except that they lack the orange color of the enamel. Immediately behind each of these incisors, and applied closely to them, is to be seen a small cylindrical tooth, the second pair of incisors. In the very young state a third pair can usually be found imbedded in the gum external to the two median ones, which fall out soon after birth. The single pair of the lower jaw are not grooved and have the usual form common to the order.

The molars are remarkable for their great length in a vertical direction, as well as their antero-posterior compression; they grow continuously and do not form roots. With the exception of the first premolar and the last molar, the molars and premolars of the upper jaw are alike, and consist of two vertical transverse laminæ closely united in the middle line, the division of which is indicated both on the inner and outer sides of the tooth by a vertical groove. The first premolar and last molar are made up of a single lamina, the enamel being thrown into two vertical folds upon the anterior part of the first premolar. In other respects the rabbits are remarkable for the entire absence of the coronoid process and the very small bony palate, which forms little more than a bridge across the roof of the mouth.

TEETH OF THE UNGULATE SERIES.

So far, excluding the rodents, our attention has been confined to those dental organs in which the molars have not, with few exceptions, passed beyond the quadritubercular stage of development; this condition, we have the best of reasons to conclude, was preceded by the trituberculate in the upper and the tuberculo-sectorial, or at least a tooth possessing its elements, in the lower jaw. When one compares these short-crowned rooted tubercular molars with the complex rootless molars of a horse, cow, or elephant, he might spend hours and days in thoughtful contemplation without discovering the faintest relationship existing between their respective patterns; nor would we be any nearer a solution of the difficulty had not the researches of palæontologists brought to our understanding a knowledge of these organs before they had assumed those distinctive characteristics and specialized patterns which they now display.

There are few students of odontography who are acquainted with the facts of mammalian palæontology as they now stand who have not had repeatedly forced upon their attention the gradual decrease in complexity of the molar teeth of the ungulates as we go backward in time. Cope has recently shown that the earliest ungulates had, as a general rule, tritubercular molars—a condition which is as primitive as that of many insectivores; and in no instance do we meet with highly specialized teeth until the latest geological periods are reached.

The ungulate series is divisible into four orders, which have been characterized and defined by Cope upon the structure of the limbs. The oldest of these orders, *Taxeopoda*, is remarkable for the generalized character of the limbs as compared with the later ungulates; they possess five toes upon each foot, and in one family, the *Periptychidæ*, the superior true molars are *tritubercular*—a fact which brings the ungulate stem to a point not far removed from the Insectivora of the unguiculates. This order includes three sub-orders, two of which are extinct, and one, the *Hyracoidea*, being represented by two living genera, popularly known as the coneys. The most ancient of these three sub-orders is the *Condylarthra*, a group thus far known only from the American Eocene. A careful study of their osteology leads to the conclusion that they are the ancestors of all succeeding ungulates, furnishing just such a generalized type in the proper geological position as is necessary to satisfy the demands of the development hypothesis; they likewise enable us to comprehend more clearly the mutual relationship and evolution of the entire series.

TEETH OF THE TAXEOPODA.—As the *Condylarthra* are the oldest of this order and the most primitive in their organization, it will be best

FIG. 250.

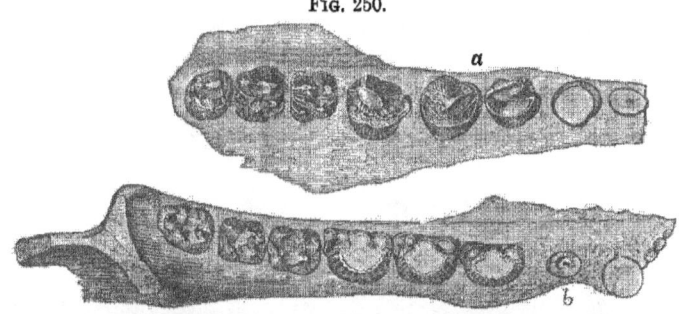

Dentition of *Periptychus rhabdodon*, Cope, two-thirds natural size: *a*, superior molars from below; *b*, inferior molars from above—from the New Mexican Puerco (after Cope).

to commence with a consideration of their teeth. Three families are referred to it, one of which, the *Periptychidæ*, is confined to the lowest Eocene deposits.[1] In the typical genus, *Periptychus* (Fig. 250), the

---

[1] When the lowest Eocene is mentioned, reference is made to the Puerco beds, which were formerly considered to belong to the Tertiary; Prof. Cope now considers that they are of Cretaceous age.

dental formula is I. $\frac{3}{3}$, C. $\frac{1}{1}$, Pm. $\frac{4}{4}$, M. $\frac{3}{3} = 44$, the normal diphyodont number. The incisors are relatively small and of the usual pattern; the canines are large, powerful teeth, and resemble those of many carnivorous and insectivorous animals. The premolars gradually increase in size from the first to the fourth, which considerably exceeds the true molars in size; the crowns of the last three premolars in the upper jaw have a large external conical cusp and a strong internal ledge; those of the lower jaw have a strong outer cusp, with a small accessory one at the antero-internal, and two at the postero-internal, angle of the crown.

The true molars of the upper series appear at first sight to be complex and multicuspid, but upon analysis it is found that they are essentially trituberculate, with minor cusps added. The two usual external cusps are present, together with one large internal tubercle somewhat crescentic in horizontal transverse section. The three principal cusps are homologous with the three cusps of the molar teeth of many of the Insectivora already mentioned, and like them are placed in the form of a triangle, but the two horns of the crescent are interrupted by the development of two intermediate cusps; to these are added two small interior cingular marginal cusps, making seven in all. The lower molars are quadrituberculate, with a faint representation of the anterior basal cusp of the tuberculo-sectorial still remaining. The postero-external cusp is connected with the antero-internal by a ridge which crosses the face of the crown obliquely; this ridge is found in some of the insectivores, notably *Esthonyx*, and is what remains of the former connection of the heel with the anterior or triangular part of the tuberculo-sectorial. The enamel of both the molars and premolars of this genus is curiously sculptured, owing to the presence of a number of

Fig. 251.

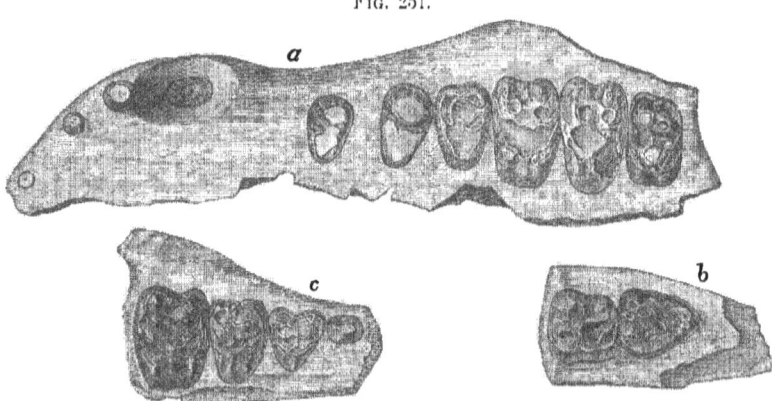

Ectoconus ditrigonus, Cope, two-thirds natural size: *a*, maxillary and premaxillary bones from below, retaining a good deal of the matrix; *b*, last two inferior molars, worn by use; *c*, three deciduous with first permanent molar of a young animal (after Cope).

vertical grooves and ridges, it being the only case of the kind known in the Mammalia. In an allied genus, *Ectoganus* (Fig. 251), the molars are larger than the premolars, and their crowns are further complicated by the addition of an outer cingular cusp, giving a total of eight of the most complex tritubercular teeth yet known. This figure displays more clearly than that of *Periptychus* the relationship of the component cusps.

Other genera of this family, of which there are seven in all, display simple tritubercular molars, which resemble the corresponding teeth of the insectivores to a remarkable extent.

The second family of this sub-order is the *Phenacodontidæ*, which continues to the Upper Eocene Period. Fragmentary remains of the typical genus *Phenacodus* were known as long ago as 1873, but very little was known of its true nature until, some nine years later, the writer was fortunate enough to discover two almost complete skeletons, representing two distinct species, in a fine state of preservation while exploring the Wasatch deposits of the Big Horn Basin, Wyoming Territory. This material has afforded Prof. Cope, at whose instance the exploration was undertaken, the opportunity of not only determining the position and affinities of this remarkable genus, but a key to a correct interpretation of many of his later discoveries, as well as a basis for one of the most important generalizations yet introduced in relation to the hoofed Mammalia.[1]

The dentition of this genus (Fig. 252) approaches nearer to that of the higher ungulates than the preceding family, although the interval between them is comparatively small. Its formula is I. $\frac{3}{3}$, C. $\frac{1}{1}$, Pm. $\frac{4}{4}$, M. $\frac{3}{3}$ = 44. The premolars are of a simpler pattern than the molars, the posterior ones becoming tritubercular. The superior molars have quadrate crowns bearing four principal cusps, placed at each angle, to which are added several minor cusps, the rudiments of structures which assume considerable importance in the later and more specialized genera. The four principal cusps are the usual ones of the quadritubercular molar, two external and two internal, and are low, more or less conic, obtuse structures. Between the outer and inner ones are two isolated tubercles, which are later developed into cross-ridges connecting the outer and inner cusps, thereby producing the lophodont molar which is so characteristic of some groups of the ungulates. At a point midway between the two outer cusps, on the external margin of the crown, the cingulum is produced into a small tubercle, which in most of the specialized ungulates becomes connected with and unites the two Vs formed by the crescentic structure of the two external cusps, just as in some of the insectivorous genera already described.

In the lower molars four tubercles are present, of which the postero-external is connected with the antero-internal by a well-marked ridge. The anterior basal lobe is reduced, but still present in the form of a low cingular ridge.

The molar teeth of this animal display a typical bunodont dentition, and upon a correct understanding of their organization depends a proper comprehension of all the succeeding specialized molars of this series. It is by simple additions to, and modifications of, the component lobes and crests of this pattern that all the complex ungulate molars have been produced; if the advocate of the evolution hypothesis had no other evi-

[1] See Prof. Cope's valuable memoir of this group, *American Naturalist* for August and September, 1884.

TEETH OF THE VERTEBRATA. 473

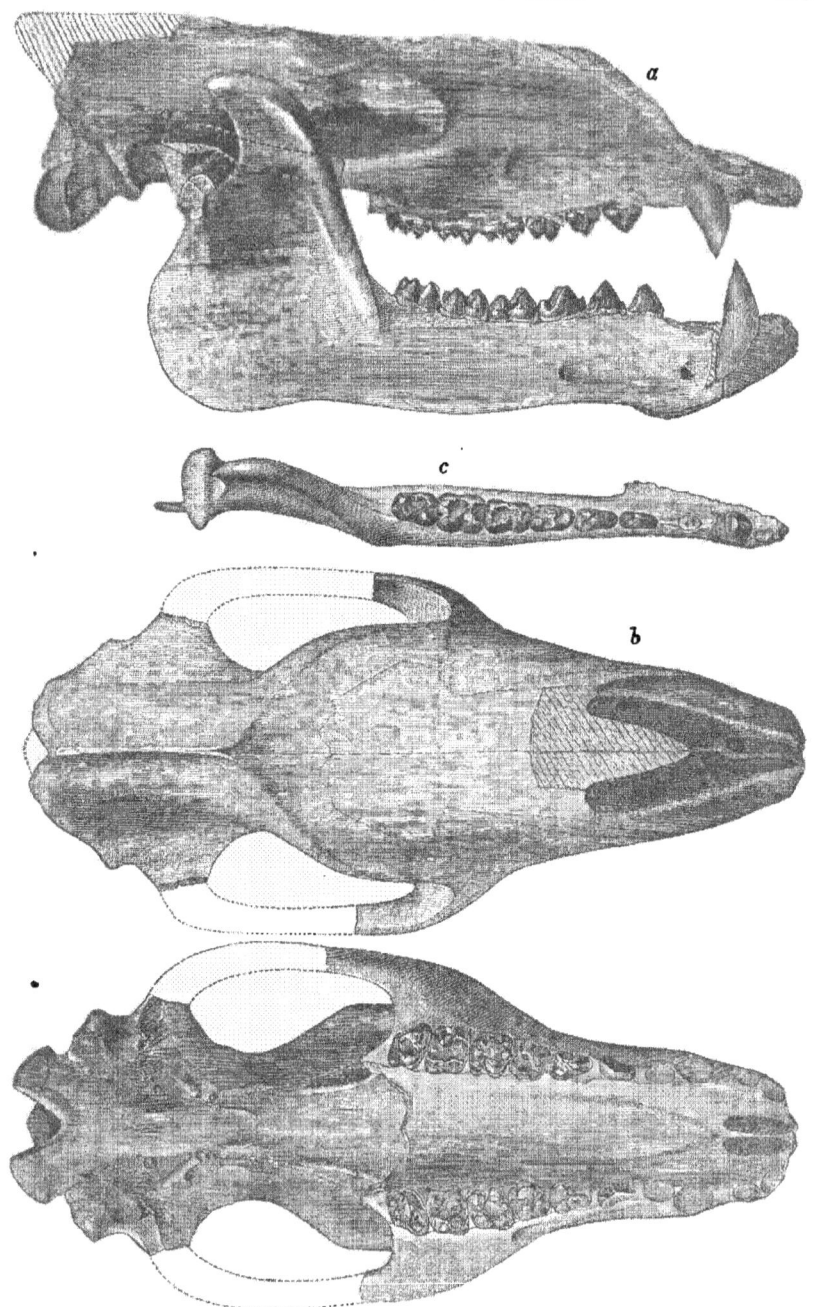

FIG. 252.—Skull of *Phenacodus primævus*, Cope one-half natural size (after Cope).

dence upon which to base his belief than that afforded by the gradual complication of the molar teeth from this point upward in the hoofed

Fig. 253.

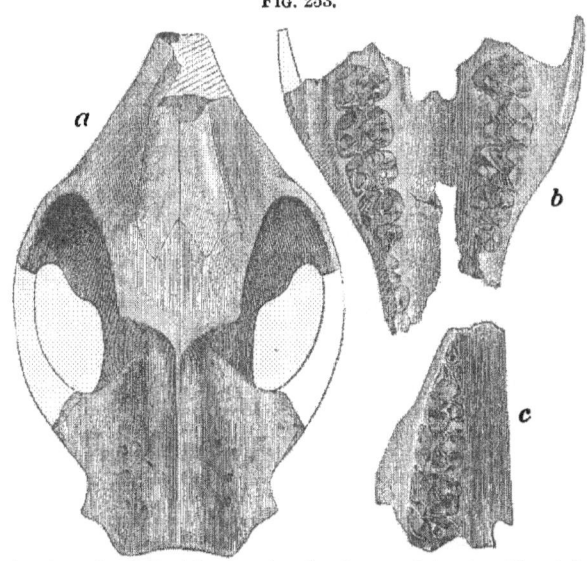

Meniscotherium terrærubræ, Parts of Cranium, three-fourths natural size—from Wasatch Beds of New Mexico: *a*, cranium from above; *b*, from below; *c*, portion of upper jaw, displaying deciduous molars (after Cope).

Mammalia, this alone, it appears to me, would be sufficient to gain for it a respectful consideration at the hands of its opponents.

Fig. 254.

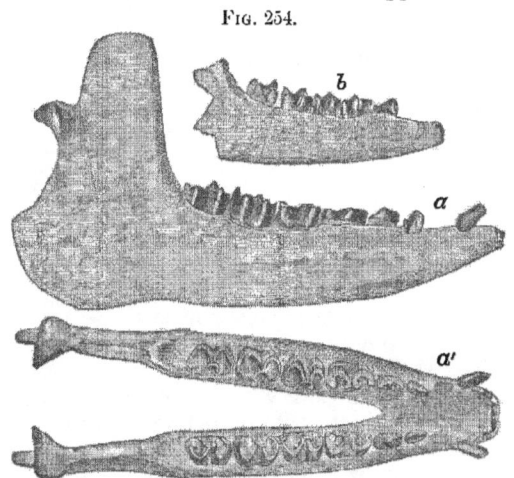

Lower Jaw of *M. terrærubræ*, three views (after Cope).

The last family of this sub-order is the *Meniscotheriidæ*, whose dentition is represented in Figs. 253, 254. The dental formula of the single genus *Meniscotherium* is given by Cope as follows: I. $\frac{3}{3}$, C. $\frac{1}{1}$, Pm. $\frac{4}{4}$, M. $\frac{3}{3}$ = 44. As compared with *Phenacodus*, the canines are relatively smaller and the molars more complex; the same elements are readily recognized as in the molars of that genus, but the two external cusps are crescentic and elevated, the two contiguous horns being connected with the median external cusp, which now forms a vertical ridge or rib on the external part of the crown. The intermediate tubercles are also present, and are greatly enlarged; the anterior is crescentic and the posterior oblique and elongate. Of the two internal cusps, the anterior is conic, while the posterior is crescentic.

The lower molars exhibit two Vs, by reason of the development of cross-ridges connecting the external with the internal cusps and the increase in height of the oblique ridge. The tooth-line is uninterrupted by a diastema, and the incisors did not grow from persistent pulps.

The second sub-order, *Hyracoidea*, has long remained a puzzle to zoologists, and has been associated at different times near the rodents, at others with the perissodactyle ungulates, and latterly has been made the type of a distinct order. The discovery of the *Condylarthra* leaves no doubt of its relationship with these forms, and the propriety of making it a sub-order of the *Taxeopoda* is at once apparent. The dental formula of the two living genera, *Hyrax* and *Dendrohyrax*, is given, I. $\frac{2}{2}$, C. $\frac{0}{0}$, Pm. $\frac{4}{4}$, M. $\frac{3}{3}$ = 36, although DeBlainville in his figures of the different species represents some of them with only two incisors in the upper jaw instead of four.

The incisors grow from persistent pulps and have large pointed crowns; the canines are entirely absent from both jaws; the molars and premolars have complex crowns—in one genus, *Hyrax*, being almost identical with those of the rhinoceros; on account of this complexity Cuvier placed it in the same group with that animal. In the other genus, *Dendrohyrax*, the molar teeth are quite different, and upon careful comparison with those of *Meniscotherium* betray unmistakable evidence of near relationship. This resemblance is not confined to the molar teeth alone, but is strikingly shown in the general form of the skull, and especially in the great enlargement of the angular portion of the mandible. The upper molars of this genus (Fig. 255),[1] like *Meniscotherium*, have two crescentic external cusps connected by a vertical rib, and two intermediate tubercles, which are more or less blended with the two internal cusps, while the lower molars have essentially the same pattern as those of this genus. The whole structure of the molars represents just such an advance over that of the extinct Eocene genus as we should be led to anticipate on *a priori* grounds.

Fig. 255.

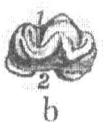

Molar Teeth of *Dendrohyrax arboreus*, vertical view: *a*, superior molar: 1, external, 2, anterior, 3, internal, and 4, posterior surfaces of crown; *b*, Inferior molar; 1, external, 2, internal surfaces (after De Blainville).

It is true that the canines are absent in *Dendrohyrax*, and incisors grow from persistent pulps; but this is not at all remarkable when we consider the great interval of time between them and an approach to this condition, as far as the canines are concerned, in their reduced size,

---

[1] This figure does not represent the structure of the grinding face very clearly.

in the extinct genus. Altogether, I am disposed to regard *Meniscotherium* as the direct ancestor of the *Hyracoidea*, notwithstanding their wide separation in both time and space.

As a further complication in the molar pattern of this line, we have the complete fusion of the intermediate tubercles with the internal cusps in *Hyrax*, which, as already stated, gives the pattern of the molars of the rhinoceros.

It is believed by Cope, from evidence afforded by the structure of the limbs, that the *Toxodontia*, a group of curious extinct ungulate forms found in the later geological horizons of South America, belong to this order. I am unable to find any confirmation of this position from a study of the teeth, but it may be that they have been derived from a condylarthrous source.[1]

The dentition of the typical genus *Toxodon* contains incisors, premolars, and molars only, the canines being absent, and all were of persistent growth. The two pairs of incisors above are large and sculpriform, as in the rodents, of which the outer greatly exceed the mesial pair in size. In the lower jaw these teeth are three in number upon each side, and were also of persistent growth. They are subequal in size, and have imperfectly prismatic crowns similar in shape to the tusks of the boar in transverse section, being covered with enamel only upon the anterior convex surface.

The molars, of which there are seven upon each side above, gradually increase in size from the first to the last. It is highly probable that the first four of these teeth are premolars, but in the absence of any knowledge of the milk dentition and the manner of its replacement, this, of course, is inferential. They have remarkably long crowns, with an altogether unique pattern, and did not develop roots. In section they are triangular, with the apex of the triangle directed forward and outward. Upon the inner side there is a deep indentation or fold reaching to a point near the centre of the crown, which may be the valley separating the two internal cusps. The only arrangement similar to this is seen in the last upper molars of many of the Ungulata, of which *Meniscotherium* furnishes an average example. Here the postero-internal cusp is absent, and the two outer cusps are intimately blended. In the rhinoceros (Fig. 257) the last molar goes even further in this direction by reason of the fusion of the elements and the obliteration of the external rib. It is conceivable that some such structure as this preceded the present one in *Toxodon*, but until the palæontological evidence of the philogeny of this group is more fully known this is the only explanation which can now be offered to account for their aberrant pattern. The pattern of the lower molars is very like that of *Meniscotherium*—a fact which lends countenance to the above hypothesis.

TEETH OF THE TRUE UNGULATA.—This order includes nearly all the modern and many extinct ungulate animals, and is conspicuously distinguished from all other hoofed forms by the interlocking character of the proximal and distal rows of the carpal and tarsal bones. Cope has called it the *Diplarthra*, in allusion to the double articular surface afforded by the ankle-bone (astragalus) to the cuboid and navicular below,

---

[1] I have elsewhere spoken of their relationship to the *Tillodontia*.

whereas in the *Condylarthra* the astragalus articulates distally with the navicular or scaphoid only—a condition which obtains in nearly all Mammalia. Two prominent divisions of this order can be recognized—the *Artiodactyla*, or "split hoofs," of which the hog, cow, and deer, etc. are familiar examples, and the *Perissodactyla*, whose only living representatives are the horse, tapir, and rhinoceros.

The latter sub-order is divisible into a number of sections, which, when we consider the extinct forms constituting at least nine-tenths of the species, we are not able to separate by any characters of very great anatomical importance, notwithstanding the fact that the extremes of the several stems are different enough. That family which stands nearest to the *Condylarthra* is the *Lophiodontidae*, a group of extinct generalized perissodactyls from the Middle and Lower Eocene beds. The digital formula is not so great as in the *Condylarthra*, being only 4—3, and in one instance 3—3; that is to say, four toes on the anterior and three on the posterior limbs.[1]

Hyracotherium (Fig. 256) is a typical example of this family, or at

FIG. 256.

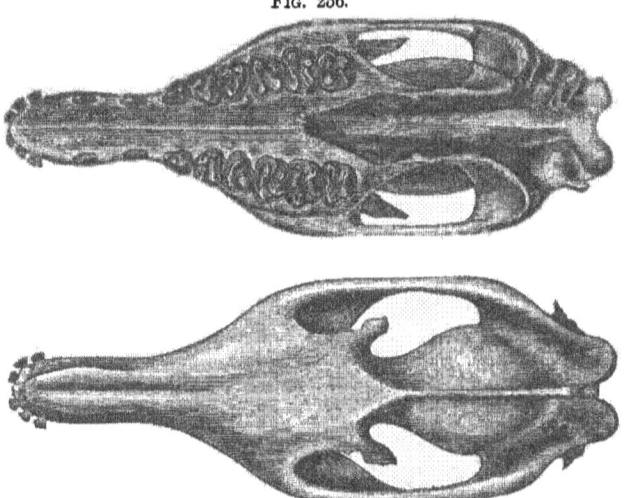

Skull of *Hyracotherium augustidens*, Cope, from the Wind River Beds of Wyoming (after Cope).

least that section of it whose dentition approaches nearest to *Phenacodus*; and if it were not known that the carpal and tarsal articulations were different, they might easily be mistaken for the same family, so great is the resemblance of their teeth. The dental formula of this animal is I. $\frac{3}{3}$, C. $\frac{1}{1}$, Pm. $\frac{4}{4}$, M. $\frac{3}{3} = 44$, the same as *Phenacodus*. The premolars are different from the molars, being simpler in form, and the first in both jaws is separated from the others by a diastema. The molar pat-

---

[1] Prof. Marsh has described several genera of this group, which he has called *Eohippus*, *Orohippus*, etc., but the descriptions are so brief that it is impossible to form any correct estimate of their true relationship. *Eohippus*, he says, has five toes, but further than this its osteology has not been described. It would be interesting to know in what respects it differs from the phenacodonts, *Hyracotherium*, *Pliolophus*, etc.

tern is substantially the same as in *Phenacodus*, with the slight exception that the cusps are more elevated and laterally flattened, and the external rib is very small or absent in *Hyracotherium*. In the nearly allied genus *Pliolophus*, which I suspect to be the same as *Orohippus* of Marsh, the last or fourth premolar below is like the true molars in form, and is quadritubercular, while the genus *Lophiotherium* has the third and fourth premolars below, like the true molars.

In the second section of this family the external lobes of the superior molars are laterally flattened and intimately blended together, so as not to be well distinguished. Of these the anterior is much the smaller, and is convex externally, whereas the posterior is large and concave without. The intermediate tubercles no longer exist as such, but form prominent crests which connect the external with the internal cusps, crossing the crown somewhat obliquely. In the lower molars the external and internal cusps are also connected by crests, giving the typical lophodont pattern. As a rule, the premolars are trilobed, and the molar formula is Pm. $\frac{4}{4}$, M. $\frac{3}{3}$, but in one genus (*Dilophodon*), recently described by Prof. Scott, there are only three premolars in the lower jaw. In another genus, lately described by the same author under the name of *Desmatotherium*, the third and fourth upper premolars are like the molars, and are four-lobed.

The tapirs form another nearly related family (*Tapiridæ*), which no doubt sprang from some member of the preceding group. The incisors and canines are like those of the *Lophiodontidæ*, but the canines in the lower jaw of the living forms are somewhat procumbent. The third and fourth premolars in the upper jaw are like the true molars, which display the four cusps connected by cross-ridges remarkable for their transverse direction in contrast with the oblique crests of some of the preceding family. The two external lobes are likewise different in their subequal proportions, both being convex externally and well separated from each other.

The lower premolars except the first are like the molars. The external and internal lobes are connected by strong cross-crests, which are as much elevated as the cusps themselves, and there is no ridge crossing from the postero-external to the antero-internal lobe, as in *Hyracotherium* and *Phenacodus*.

From this family we pass to the rhinoceros section of the sub-order. In accordance with what the philosophic student of the living forms would be led to anticipate, this section pertains to a later geologic period than the preceding, and not unnaturally would he seek for the connecting links between them and that section of the *Lophiodontidæ*, in which the external lobes are flattened. Through the researches of American palæontologists we are now in a position to fully comprehend all the more important steps in the evolution of this group, and I fail to recall in the whole range of vertebrate palæontology an instance in which the demands of the evolution hypothesis are more completely satisfied than in the present one.

The molar formula of the rhinoceros is Pm. $\frac{4}{4}$, M. $\frac{3}{3}$, the usual number in perissodactyles; but, as regards the incisors and canines, the greatest variability is to be observed. In the two-horned African species neither canines nor incisors exist in the adult animal, they having completely disappeared in the course of development. On the other hand, in the remarkable and interesting Eocene genus *Orthocynodon* of Scott and Osborn,[1] the canines are of normal size and erect in position, as the name implies. The number of incisors has not been definitely determined, owing to the imperfect condition of the single specimen

Fig. 257.

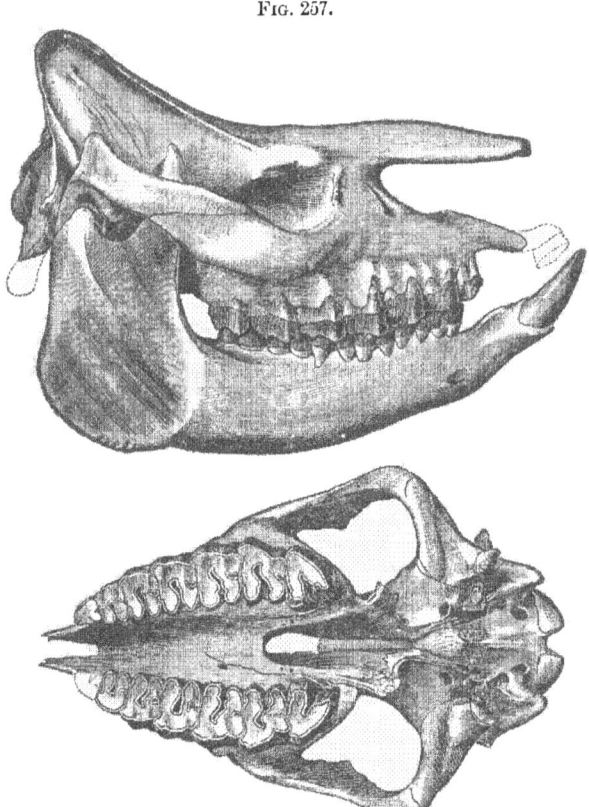

Skull of *Aphelops megalodus*, Cope, an extinct American rhinoceros.

known, but it certainly had two, and probably three, in the lower jaw, as there is abundance of room between the canine and the second incisor for another tooth. If the number is three in the lower jaw, it would imply a like number in the upper, which would bring it very near to the *Lophiodontidæ*. From this condition of the dentition, which is very nearly that of the *Lophiodontidæ*, we pass to the genus *Amynodon* of Prof. Marsh, in which the lower canine is much smaller and procumbent, with the incisors reduced to two pairs in each jaw. Following this

[1] See *Bulletin No. 3 Contributions* from E. M. Museum of Geology and Archæology of Princeton College, May, 1883.

genus in time comes the Lower Miocene representative *Aceratherium*, in which the incisors are two upon each side in the upper and one in the lower jaw, with the upper canine absent. The Middle Miocene furnishes a genus, *Ceratorhinus*, in which the incisors are one upon each side above and below, and a canine in the lower jaw only. Finally, we have a complete disappearance of both incisors and canines in some species now living. The reduction of the incisors and canines from *Orthocynodon* to *Cœlodonta*, a living species, can be summarized as follows: I. $\frac{3}{3}$, C. $\frac{1}{1}$, *Orthocynodon* ; I. $\frac{2}{2}$, C. $\frac{1}{1}$, *Amynodon* ; I. $\frac{2}{1}$, C. $\frac{0}{1}$, *Aceratherium* ; I. $\frac{1}{1}$, C. $\frac{0}{1}$, *Ceratorhinus* ; I. $\frac{0}{0}$, C. $\frac{0}{0}$, *Cœlodonta*.

Fig. 258.

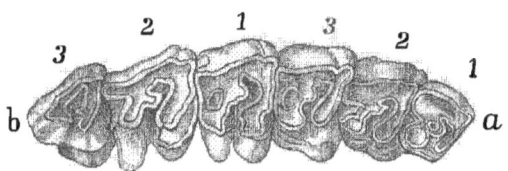

Superior Molar Dentition of Rhinoceros: *a*, anterior; *b*, posterior end of series. The figures 1, 2, 3 indicate molars and premolars.

In the earliest forms the molars are more complex than the premolars, but in the later and living species the premolars are as highly organized as the molars, and like them in form; this is well shown in the accompanying figure.

About twelve genera have been described, seven of which come from the fossil beds of North America. Through *Orthocynodon*, as was pointed out by Profs. Scott and Osborn, they inosculate with the *Lophiodontidæ*, after which they branch into several distinct lines. While the rhinoceroses have perpetuated the type of molar which began with the last section of the lophiodonts, other forms inherited the pattern of the hyracotheroids, and from this point the dentition was gradually specialized, not so much through subtraction of the number of teeth as addition and complication of the different lobes and crests of the crowns of both molars and premolars. The culminating point of this line is found in the living horses.

The first step beyond *Hyracotherium* in this series is seen in the Eocene genus *Ectocium* of Cope, in which the external rib is better defined, the external cusps more crescentic, and the cusps and oblique ridge of the lower molars are more prominent. In the next geological stage (Upper Eocene) we meet with the family *Chalicotheriidæ*, abundantly represented in the Wind River deposits by the genus *Lambdotherium*, likewise described by Cope. In this form (Fig. 259) the external cusps of the upper molars are considerably elevated, of a crescentic form, and connected with an external median rib. The anterior cross-crest still has a tubercular form, while the posterior is crest-like and blended with the postero-internal lobe. The four cusps of the inferior molars are connected by cross-ridges and the oblique crest, so as to form two Vs, opening internally. The tooth here represented is the last one, which in many of the perissodactyls has a prominent heel (*h*). In this animal the antero-internal cusp (*ai*) becomes bifid at its summit, and the anterior basal lobe (*k*) again assumes considerable importance.

Following the chalicotherioids, and as a probable derivative of them, we meet with the palæotherioids, in which the molar pattern makes a considerable advance in complexity over that of the preceding family, and the premolars are now like the molars. *Anchitherium* is a good representative of this group, and is here taken for illustration. This genus is of especial interest, in view of its ancestral relation with the horses; it is here that we get the first distinctive traces of equine peculiarities, although several genera intervene between it and the modern Equidæ or horses. The species were numerous, most of them equalling the sheep in size, and had three subequal toes on each foot.

FIG. 259.

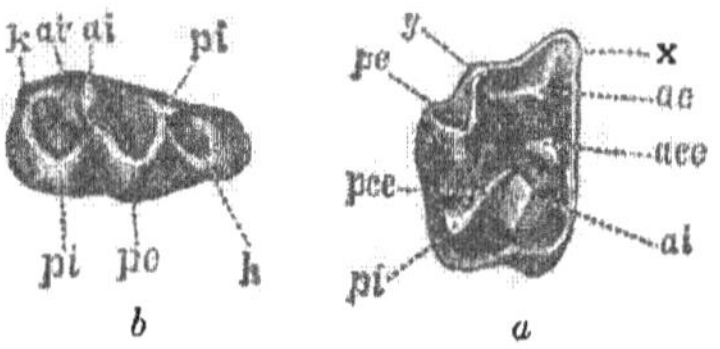

Upper and Lower Molar Teeth of *Lambdotherium*, vertical view, natural size: *a*, superior; *b*, last inferior molar. In the upper molar, *ae*, antero-external; *pe*, postero-external; *ai*, antero-internal; *pi*, postero-internal or principal cusps respectively: *y*, external vertical rib; *x*, an anterior cingular cusp; *ace-pce*, anterior and posterior cross-crests. In the lower molar the principal cusps are lettered the same: *k*, anterior basal lobe; *ai'*, accessory cusp; *h*, heel.

The incisors, as in all the preceding genera, are plain incisiform teeth, without the pits or "mark" found in the corresponding teeth of the horses. Well-developed canines are likewise present.

The superior molars (Fig. 260) display the same elements as those of *Lambdotherium*; the external cusps are very much flattened and crescentic, having their vertical dimensions considerably augmented. The cross-crests form laminar ridges connected with the two internal cusps at the base, and separated from them above by open notches; they reach quite across the face of the crown. The two internal cusps almost equal the external ones in height, but have a more conical form; they are separated from each other by a deep fissure or valley opening internally. On the posterior border of the crown the cingulum develops an accessory cusp, which has a tendency to form a cross-crest in this situation and enclose a valley between it and the posterior cross-crest.

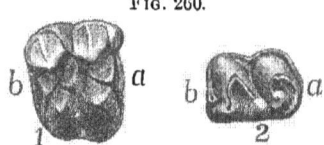

FIG. 260.

Upper and Lower Molars of Right Side of a species of *Anchitherium*: 1, upper; 2, lower tooth; *a*, anterior; *b*, posterior border.

In the molars of the lower jaw the same elevation of the lobes and crests is to be observed; their pattern is substantially that of *Lambdotherium*.

In a later geological epoch the genus *Hippotherium* carries dental modification a step farther toward that of the existing horses. The outer toes are much reduced, the incisors possess the peculiar pits of the horse, the molars are more complicated, and the entire appearance is decidedly equine. A strict comparison of the elements of the molars with those of *Anchitherium* is generally difficult, on account of the thick deposit of cement which fills up the valleys and spaces between them. To obviate this difficulty and bring out more clearly the relationship between them, I have represented in Fig. 261 an unworn molar in which the cementum has been removed. Although the respective patterns are very much alike in their general structure, the differences consist in this: the external cusps of the superior molars are relatively larger, more perfectly crescentic, and strongly inclined inward in *Hippotherium*. The anterior cross-crest is better developed and joins the posterior cross-crest, so as to enclose a deep pit or valley between it and the antero-external cusp, which is filled with cement in the natural

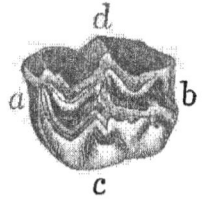

Fig. 261.

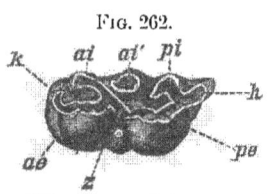

Fig. 262.

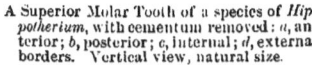

A Superior Molar Tooth of a species of *Hippotherium*, with cementum removed: *a*, anterior; *b*, posterior; *c*, internal; *d*, external borders. Vertical view, natural size.

Lower Molar of same. Letters as in Fig. 259.

state; this is called the anterior lake in the worn tooth. The posterior cross-crest bends around to join the posterior cingular cusp, which, with the postero-external cusp, furnishes the boundary of the posterior lake. To these cross-crests are added a greater or lesser number of vertical folds, which give the borders of the lakes a crenate appearance when the crown is much worn. The internal cusps are relatively small, the posterior being connected with the corresponding cross-crest, the anterior isolated. To all these must be added the increased height and the presence of cementum.

The lower molars (Fig. 262) do not exhibit such marked difference from the *Anchitherium* type as do those above, but they are nevertheless more complex in their increased depth, complete isolation of the accessory antero-internal cusp, and the addition of cementum. The grinding surface of the teeth resulting from this arrangement of the enamel, dentine, and cement is kept constantly rough by reason of the inequalities in the rate of wear which these substances sustain during mastication. Coincident with this structure of the crown the roots disappear and the tooth grows continuously—a condition necessary to compensate for the great waste of the tooth-substance.

Lastly, we come to the modern horse, in which digital reduction has reached the extreme point in this series, or that furthest removed from the pentedactyle *Condylarthra*. As is well known, the digital formula in this family is 1 − 1 in functional use, with the second and fourth represented by the rudimentary metapodials commonly known as the "splint bones."

The incisors are peculiar and characteristic, inasmuch as the working face is interrupted by a deep pit caused by the upward growth of the posterior cingulum. Previous to extrusion, the posterior wall of this cavity is incomplete and does not rise so high as the anterior.[1] After

---

[1] Ryder, "On the Origin and Homologies of the Incisors of the Horse," *Proc. Acad. Nat. Sci.*, Philada., 1877.

the tooth has been in use for a little while, however, the face is worn down smooth, and the central depression appears bounded by a layer of enamel, between which and the enamel covering the outer surface of

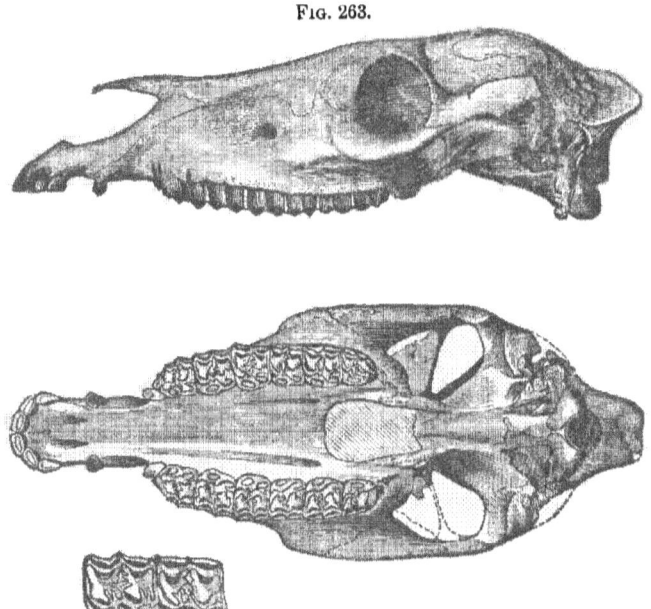

Fig. 263.

Skull of *Hippotherium severum*, Cope (after Cope).

the tooth, may be seen the dentine. The incisors are not all cut at the same time, the last appearing at the age of five years, on account of which the central pit disappears through wear sooner in those teeth which are first extruded than those which are cut last. By observing carefully the date of appearance of the various incisors, and the consequent difference in time at which the pits are obliterated in the different teeth, veterinarians have established some very useful rules by which the age of a horse can be approximately told with considerable certainty up to ten or twelve years.

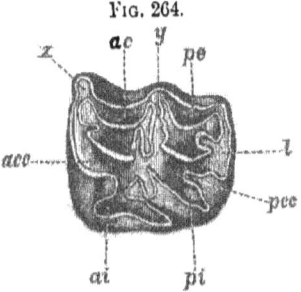

Fig. 264.

Molar Tooth of a species of Horse. Letters as in the preceding figures.

Canines, or the "bridle teeth," are present, but they are of smaller size, and sometimes disappear in the female. The first premolars in both jaws are normally absent, but there are many cases on record in which they are present. In *Hippotherium* they are normally present and functional.

The molars present essentially the same pattern as those of the preceding genus, the only difference of importance being found in the enlargement of the antero-internal lobe and its connection by a ridge with its corresponding cross-crest. Some species of *Hippotherium* show a gradual advance from the conic isolated condition of this element to its enlarged and sub-connected form.

Thus it is that palæontology has enabled us to fully comprehend the different steps in the production of these complex and specialized organs from the simple bunodont pattern. To say that such evidence is without its special bearing on the great problem of biology, or that evolution or development has not taken place, is to deny the truth of the assertions herein made. Many intermediate steps between those given could be cited, but time and space have compelled me to limit the examples to the most salient.

The remaining perissodactyles exhibit different degrees of modification of the bunodont type, none having reached the same stage of perfection as the horse.

The second sub-order of the ungulates, *Artiodactyla*, attained its greatest development at a later geological period, and it is probably in the present epoch that the genera and species are the most numerous. A few genera are found in the Lower Eocene, but they are of rare occurrence as compared with the perissodactyles. It is probable that they two came off the condylarthrous stem, but the direct evidence to substantiate this supposition is wanting. They are primarily divisible into two groups, *Bunodontia* and *Selenodontia*, characterized by the pattern of the molar teeth and the consequent condition of the posterior termination of the maxillary bones. In the former division, of which the hog is an excellent example, the molars have approximately the same pattern as *Phenacodus;* the tooth-line is little curved, and the posterior extremity of the maxillary is applied closely to the palatine and pterygoid bones, whereas in the *Selenodontia* the molar teeth have crescentic cusps, and the posterior borders of the maxillaries are separated by a wide sinus from the palatines and pterygoids. These characters at first appear insignificant and inadequate to establish and define two such great groups as the foregoing; but when we remember that they express a very important structural modification, and that the two are correlated, we cease to express surprise.

Of these two divisions, the *Bunodontia* is the older, and as a consequence the more generalized. Their generalized characters are most conspicuously displayed in the increased number of digits, bunodont teeth, absence of horns, non-complexity of the stomach, and separate condition of all the limb bones. In fact, the suilline artiodactyles are as primitive in many respects as the *Condylarthra*, but in the arrangement of the carpal and tarsal elements they are specialized and far removed from their primitive ancestry.

In the hog the dental formula is I. $\frac{3}{3}$, C. $\frac{1}{1}$, Pm. $\frac{4}{4}$, M. $\frac{3}{3} = 44$. The outer pair of incisors are small, and sometimes fall out in old age. The canines are relatively large—disproportionally so in the male—and in the upper jaw curve round in such a manner that the point of the crown is directed upward. The enamel of these teeth does not uniformly invest the crown, but is disposed in three bands corresponding with its trihedral form. The canines of the lower jaw are more slender and have a normal direction. It is said that castration arrests the excessive development of the tusks of the boar, just as this operation profoundly affects the growth of the antlers of the deer—a circumstance which at once relegates the cause of this condition to sexual influences.

The first premolar has no deciduous predecessor, and disappears soon after the adult stage is reached. The rest of the premolars increase in complexity and size from front to rear, but none of them are quadrituberular. The first and second molars are quadrate in section, with four-lobed crowns. The last molar is greatly elongated in an anteroposterior direction, which is occasioned by the possession of an enormous heel, much as in the bears, and its crown, as in the others, besides presenting the normal four cusps, has an immense number of subsidiary tubercles, giving to it a decidedly wrinkled appearance.

In the wart-hogs (*Phacochœrus*) a very peculiar modification of the molar pattern is to be seen in the last tooth. In the unworn state the crown of this tooth presents about thirty small tubercles, arranged in three rows in a direction longitudinal to the axis of the body, the intermediate spaces between them being occupied by cementum. When wear takes place, the summits of these cusps are abraded, leaving as many little dentine islands bordered by enamel; they are strengthened by the addition of cement.

The canines are of enormous size, devoid of enamel, and grow from persistent pulps; the superior ones are directed upward at first, piercing the upper lip, and then curve backward toward the eye; their length is sometimes as much as eight or ten inches. All the molar teeth are generally shed in old age, with the exception of the fourth premolar and the last true molar, so that the molar dentition is practically reduced at this time to four upon each side in both jaws, and is the only case of the kind known in the Mammalia.

The peccaries constitute another family of this division, and are known from the lowest Miocene, if not from the Upper Eocene deposits. Their molar dentition is more nearly like that of the *Condylarthra* and primitive perissodactyles than other suillines, lacking, as a rule, the great development of the minor tubercles of the molars of the hog as well as the elongated heel of the tooth. The canines, moreover, are normal in direction, and the great disparity in size between these teeth does not exist in the sexes. The incisors are of the usual pattern, although the outer pair is absent from both jaws in some genera.

From this family the transition is easy to the earlier forms of the selenodonts, in which the feet were multidactyle; in one genus, *Oreodon*, as has been recently shown by Prof. W. B. Scott, the anterior limb was provided with the normal number of toes, five. That family, which almost completely bridges the chasm between these divisions, is the extinct *Anthracotheridæ*, whose remains are abundant in the Miocene strata of Europe, but less so in this country.

It is somewhat uncertain how many genera should be referred to this family, and by what character or characters it should be defined. *Palœochœrus*, which by common consent is a suilline, has four lobes upon the crowns of the superior molars, which are conic and not connected with an external rib, together with two small intermediate cusps, representing the cross-crests very much as in *Phenacodus*. *Chœropotamus*, another genus from the Eocene of France, is altogether intermediate between *Palæochœrus* and *Anthracotherium*, the typical representative of this family, in the pattern of the superior molars; the external cusps are somewhat crescentic, but the external rib is rudimentary or absent. In the first molar the anterior of the two intermediate tubercles only is present, while in the other two molars it is very small and insignificant; the two internal lobes are conic.

Following this genus in time come *Anthracotherium*, *Hyopotamus*,[1] *Ancodus*, and others in which the anterior of the two intermediate tubercles is the only one which is present in the upper molars. This character, I am therefore disposed to believe, defines a natural group, and should, in connection with the external rib and crescentic form of the external cusps, be the test of limitation of this family.

Two derivatives of the Eocene *Hyopotami*, *Xiphodon*, and *Anoplotherium* soon became specialized in their limb structure, but, strangely enough, disappeared in the Early Miocene. Another line was commenced contemporaneously with that of the anthracotheroid in the genus *Dichobune*, wherein the *posterior* intermediate tubercle only was retained. It continues forward through the genus *Cainotherium* into the Upper Miocene deposits of Sansan, where it gradually faded from existence, leaving no modified descendants. This, it appears to me, constitutes another family, definable by the above character.

From the *Anthracotheridæ* have sprung all the modern artiodactyles, with the possible exception of the cameloids and the existing suillines, together with other stems which are extinct. Many extinct genera complete the connections with the living forms in all the osteological and dental details, which it is scarcely within the scope of the present article to discuss.

In the production of a perfected double crescentic pattern of the superior molars in this sub-order from the short-crowned semi-bunodont anthracotheroids, the anterior intermediate tubercle has gradually usurped the function of the true antero-internal cusp, it having been reduced to a small cusp situated internal to the mesial horns of the inner crescents on the inner basal portion of the crown (see Fig. 265).[2]

Specialization of the dental organs of the *Selenodontia* is seen in the following characters: (1) Formation of double crescents in the superior and inferior molars; (2) great elevation of the cusps and deposit of a thick layer of cementum, filling up the valleys; (3) loss of the roots of the molars and premolars, and their growth from persistent pulps; (4) reduction of the premolars to three in each jaw; (5) subtraction of the canines and incisors from the upper jaw; (6) the reduction in size and approximation of the lower canine to the incisors; and finally (7), the

---

[1] Gaudry places the appearance of this genus in the sands of Beauchamp, which probably corresponds with our Bridger Beds or Upper Eocene. He also fixes the date of appearance of *Palæochœrus* in Europe in the deposits of Saint-Gerand-le-Puy, Middle Miocene. In this country *Hyopotamus* does not appear until the Lower Miocene, whereas *Palæochœrus* probably extends into the Bridger epoch.

[2] For a further knowledge of the fossil forms of these families the reader is referred to the important work of Prof. Albert Gaudry, "Les Enchainements du Monde animal dans les Temps géologiques," in which the more important genera are figured.

TEETH OF THE VERTEBRATA. 487

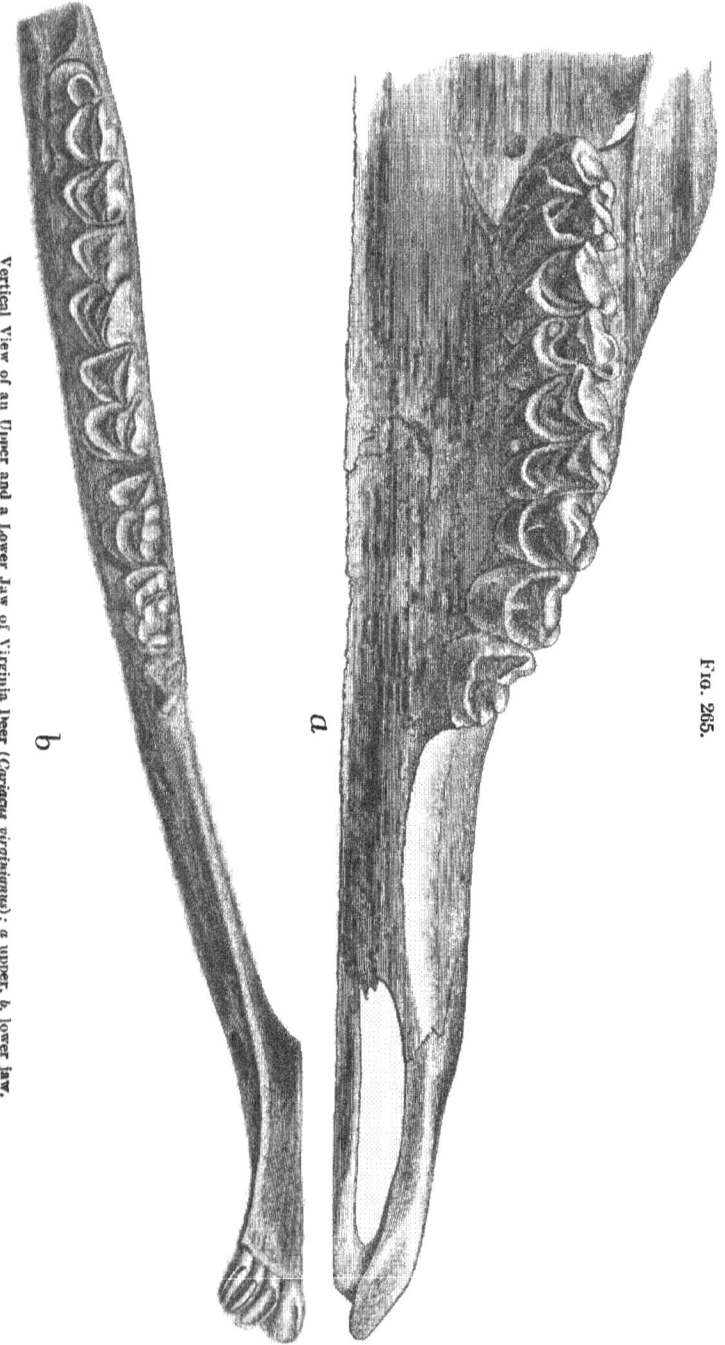

Fig. 265.

Vertical View of an Upper and a Lower Jaw of Virginia Deer (*Cariacus virginianus*): *a* upper, *b*, lower jaw.

development of a long diastema in front of the premolars. While the complete assumption of these characters is reached only in the bovine ruminants, others exhibit all the intermediate stages of modification tending in that direction.

The common Virginia deer (*Cariacus virginianus*) has been selected as an average example of the higher selenodont dentition; although in its family (Cervidæ) canines are sometimes found in the upper jaw, there is little or no cementum on the crowns of the molars, and they have well-defined roots. It will therefore be observed that it does not fulfil all the requirements of the most highly specialized selenodonts in its dental organization. The dental formula of this species (Fig. 265) is I. $\frac{0}{3}$, C. $\frac{0}{1}$, Pm. $\frac{3}{3}$, M. $\frac{3}{3}$ = 32. The incisors have long spatulate crowns, the median pair being the larger, the outer ones decreasing gradually in size. The canines are smaller than the outer pair of incisors, which they resemble very much in shape, being applied closely to them. After an immense interval follow the premolars, the first two in the lower jaw being comparatively simple, the third four-lobed like the succeeding molars. The molars display two perfect double crescents, of which the outer are convex externally. The last molar has a fifth lobe. In the upper jaw the premolars are bilobed, the internal being convex internally and enclosing a deep valley between it and the external cusp. The true molars have double crescents enclosing two valleys. The antero-internal of these crescents is made up of the anterior intermediate tubercle, which has become greatly enlarged and developed into a crescentic form, the true antero-internal cusp being situated internal to and behind it. The proper evidence to support this determination is to be found by examining the superior molars of *Hyopotamus*, *Anoplotherium*, and *Xiphodon*, it which it will be seen that the antero-internal cusp becomes gradually smaller.

Teeth of the Proboscidea.

The last order of the ungulate series whose dental organs remain to be noticed is that including the elephants, mastodons, etc. The animals composing this group are the largest of terrestrial mammals, and display many curious modifications of the primitive ungulate type. Probably no part of their organization has been more profoundly affected in their gradual evolutionary growth than the teeth, and were it not for the fact that abundant evidence is at hand to demonstrate the successive steps in the progressive modification from a more simple type, we would be at a loss to comprehend the manner of production of these most complex of all teeth.

Two genera of proboscideans are found in the existing faunæ of Asia and Africa, but these are only the inconsiderable remnant of a once greater and much more widely distributed representation, as is indicated by their fossil remains. During the later Tertiaries proboscideans were not unknown in both the northern and southern hemispheres in all the extensive land-areas; in some parts of the northern hemisphere, where they are now extinct, judging from their fossil remains immense herds and droves must have at one time existed.

In the African elephant (*Loxodon africanus*) the dental formula is I. $\frac{1}{0}$, C. $\frac{0}{0}$, Pm. and M. $\frac{6}{6}$. The two incisors are greatly enlarged, implanted in deep sockets, and grow from persistent pulps. They are preceded by small deciduous teeth, and when first protruded are tipped with enamel, which soon wears off. The tooth then consists mainly of dentine covered by a thin layer of cement, the dentine presenting a slightly modified form known as "ivory." This substance, as is well known, is extensively used in the arts and has a fixed commercial value. Although not exclusively confined to the tusks of the elephant, nevertheless the chief source of supply of this material is derived from them. Tomes cites an example in which a pair of tusks of this species were exhibited in England that weighed three hundred and twenty-five pounds and measured eight feet six inches in length and twenty-two inches in circumference; the average weight, however, does not exceed from twenty to fifty pounds. The female of this species has tusks quite as large as the male, but in the Indian species the tusks of the male exceed those of the female in size.

The molar teeth of the living elephants are very much alike in general pattern and mode of replacement, which is unique; the description of one will therefore suffice to convey an intelligent understanding of the entire subject.

Both existing species have a molar formula of $\frac{6}{6}$, which are divided into milk molars $\frac{3}{3}$, true molars $\frac{3}{3}$. There is sometimes, in addition to these, a small rudimentary milk molar in front, which increases the total number to seven upon either side in each jaw.

Although the total number of molars is normal or nearly so, they are not all in place nor in existence at the same time. Barring the occasional rudimentary one, the first molar in the Indian species cuts the gum at a considerable distance from the front of the jaw about the second week after birth. It is implanted by two fangs, and displays a subcompressed crown bearing four cross-ridges, and is therefore lophodont in pattern. The upper tooth corresponding to this one cuts the gum a little earlier, and possesses five cross-crests. These teeth are shed at about the age of two years.

Before the disappearance of the first two teeth the second molars come into place from behind. They are considerably larger than the first, being on an average two and a half inches in length by one inch in breadth. Their crowns are of similar form, but have the number of cross-ridges increased to eight or nine. They are implanted by two fangs, and are shed before the beginning of the sixth year.

By the time the second molar has been worn out the third molar, averaging four inches in length by two in breadth, makes its appearance. Its crown has from eleven to thirteen cross-plates on its working face, and is also supported by two fangs, of which the posterior is much the larger. It is said to be worn out and shed about the ninth year.

The teeth so far enumerated are taken to be homologous with the second, third, and fourth milk molars of the ordinary diphyodont dentition, which have in this case failed to develop permanent successors. This conclusion is rendered reasonably certain, as we shall presently see, by the fact that their ancestors had a more or less complete permanent premolar system, which underwent progressive subtraction as they approached the modern proboscideans.

Three teeth which are homologous with the permanent true molars are developed behind these in a similar manner. They increase in size and complexity from before backward: the first, or fourth of the entire series, bears fifteen or sixteen plates; the second has from seventeen to twenty plates; while the last supports from twenty to twenty-five. The first true molar disappears between the twentieth and twenty-fifth years of the animal's life, the second somewhere about the sixtieth, while the last is retained until the termination of the animal's natural existence, which is said to be more than one hundred years.

The structure of these teeth is complex, and, as we have said on a former page, resembles that of some of the hystricine rodents, such as the capybara, for example. The cross-ridges near their summits are broken up into a number of conical projections, which, when abrasion first takes place, present so many dentine islands surrounded by a rim of enamel: these are arranged in rows across the face of the crown in the position of the future plate (see Fig. 266). As wear goes on these islands unite

Fig. 266.

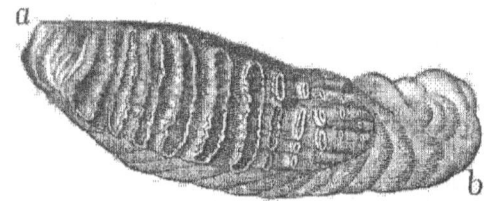

Molar Teeth of Indian Elephant (*Elephas indicus*), after Tomes: *a*, anterior; *b*, posterior border.

below, and form transverse lamellæ composed of a narrow strip of dentine surrounded by enamel. Between these much-elongated lamellæ, which are all blended together at the base of the crown, a thick deposit of cementum is found; it also invests the lateral surfaces of the crown and prevents fracture of the cross-plates.

In the growth of the tooth the anterior plates or crests are first formed, and come into position and use long before the posterior. As a consequence of this, the most anterior plates wear out and disappear while the posterior ones are still being formed. This is well shown in the accompanying figure. As new plates are added from behind, the whole tooth moves forward, which probably exerts some influence in the removal of the tooth in front of it. Finally, before the tooth disappears altogether, it presents an oval area of smooth dentine surrounded by enamel and cementum. It is then no longer efficient as a grinding organ, and is consequently discarded.

It will be seen by this arrangement of the three tooth-substances on the working surface of the crown, and by reason of the varying rate of their wear, the teeth of the two jaws when brought into opposition afford most perfect machinery for the grinding up of the coarse herbaceous substances upon which the elephant feeds.

The two genera of existing proboscideans may be readily distinguished by the character of the plates of the molar teeth. In the African species they are fewer in number on the corresponding teeth than in the Indian, and they have a distinct lozenge-shaped pattern upon cross-section, whereas in the Indian species they present an oval outline upon cross-section and the enamel border is crenate. In the number and succession of the teeth the two genera are alike.

The genus *Deinotherium* includes a few species whose remains have been found in the Miocene deposits of Europe, and which were but little if any inferior to the living proboscideans in bodily proportions. They are the oldest representatives of this order so far discovered, and especial interest attaches to their teeth, inasmuch as their structure furnishes a clue to a more perfect understanding of the later and more complex types.

The premaxillary bones were edentulous, but the front part of the lower jaw was provided with two large decurved tusks. What particular use the animal made of these teeth is difficult to imagine. The molar formula is Pm. ⅔, M. ⅔. The structure of these teeth is not very different from that of the tapir, consisting of a moderately short crown bearing two or three cross-crests. Both the premolars had deciduous predecessors, just as in the diphyodonts generally. These animals, however, were very elephantine in every other feature of their anatomy, and were in all probability provided with a trunk.

From this condition of the dental organs we pass to the mastodons, in which there is a marked approach to the elephants. In some species there were two tusks in each jaw, but the lower ones were small, and in many cases disappeared early in life. The molars increase in complexity and size from before backward, the posterior ones bearing in some species as many as ten cross-crests, which were unsupported by a cementum deposit; in others the cross-ridges are much fewer in number. Many species are known, and when all are considered a complete transition between the comparatively simple lophodont and the extreme lamellate patterns is afforded. Many of them had deciduous teeth, which were vertically succeeded by two, and probably three, permanent premolars. As the elephantine molar pattern was acquired, however, these were gradually lost.

Altogether, it is impossible for a student of odontography to study carefully the teeth of this order, and not be thoroughly convinced in the end that the complex pattern has gradually, but none the less certainly, arisen from the simpler one. If this, therefore, is true of one series, it must be of all.

The Amblypoda.

Another order of hoofed mammals which became extinct at the close of the Eocene Period has been described from the fossil-bearing deposits of this country. They were mostly of gigantic proportions, and exhibit affinities with both the proboscideans and the Perissodactyla. They are most nearly related, however, to the Toxcopoda, with which they were contemporary in the Eocene.

Nearly all of them have the full complement of incisors, canines, pre-

FIG. 267.

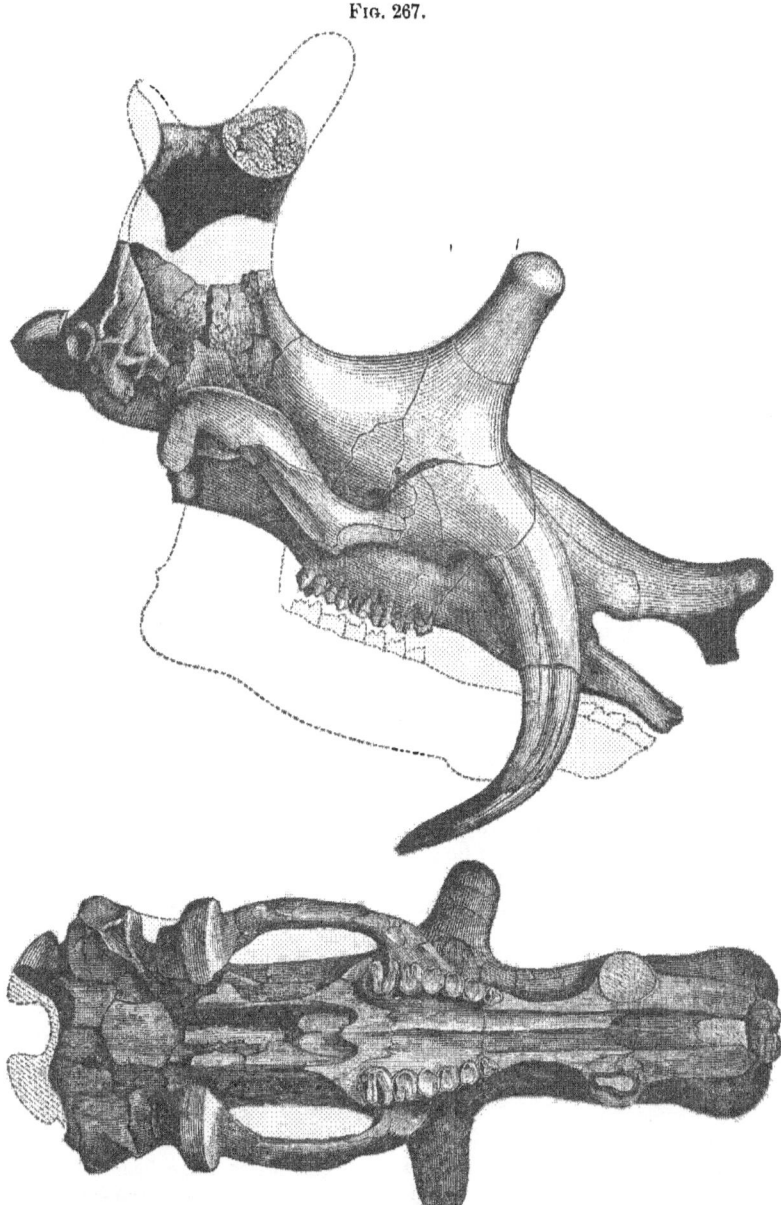

Skull of *Loxolophodon cornutus*, Cope, a species of amblypod from the American Eocene (after Cope).

molars, and molars, and in some the canines were greatly enlarged. The molar pattern is of moderate complexity, and shows a considerable departure from the primitive tritubercular ancestry. In the lower jaw the molars are lophodont, while in the upper they have a single crescent of moderate perfection. Owing to their near relationship with the Toxeopoda, it is highly probable that their teeth represent an extreme modification of the tritubercular pattern, but of the different steps in their production lack of space prevents me from speaking here. I must refer the reader to the papers of Profs. Cope and Marsh for a more complete description of the dentition of this order.

TEETH OF THE MARSUPIALS.

I have indicated on a preceding page that this division of the Mammalia is sharply defined from the monodelphs by the circumstance that no connections are formed between the foetal envelopes and the walls of the uterine cavity during gestation, so that no placenta is developed. They are therefore known as the implacental division of the *Eutheria;* they are likewise known as the *Didelphia* and *Marsupialia.* The young are born in an exceedingly helpless and imperfect condition, and are transferred to the pouch or marsupium of the mother, where, by a special arrangement, the nourishment is forced into their mouths until such time as they are enabled to help themselves.

In the majority of the lower Vertebrata very little development of the young takes place in the body-cavity of the mother; the ovum is relatively large, by reason of the addition of an abundant supply of pabulum sufficient to nourish the embryo until the later stages of development are reached. It has been recently ascertained that the monotremes reproduce in the same way; that is, they lay eggs like birds and reptiles, which are hatched in a similar manner. The whole plan of development moreover, is like that of the bird (mesoblastic)—a condition which would be reasonably suggested by a study of their reproductive system.

As the monotremes furnish the connecting link between the higher mammal and the reptile, so do the marsupials, as far as reproduction is concerned, afford a transitional stage between the monotremes and the monodelphs. For this reason we would naturally be led to look for primitive and transitional characters in their teeth. Unfortunately, these organs do not in many particulars go beyond the lowest forms of the monodelphs sufficiently to give us any clear insight into the intermediate structures and patterns which must have preceded the diphyodont monodelph dentition; still, some of the earliest representatives of mammalian existence which have been referred to in this group possess a greater number of heterodont molar and premolar teeth than any known mammal.

In the small living marsupial genus *Myrmecobius* the dental formula is I. $\frac{4}{3}$, C. $\frac{1}{1}$, Pm. $\frac{3}{3}$, M. $\frac{6}{6} = 54$. The incisors are small, subconic teeth, implanted in the premaxillary bones above, and followed by the canines, which have the usual laniary form. The premolars have laterally-compressed, unicuspid crowns, and are implanted by two roots. The molars exceed in number those of any other marsupial, reaching the unusual number of six in each jaw. Owing to the imperfect descriptions of the crowns of these teeth, and never having seen a specimen myself, I am at present unable to say just what the pattern of the crown is. From the best information at my command I suppose it to be somewhat after the style of a modified tuberculo-sectorial. I further do not know whether the succession has been observed, and whether a proper distribution of the molars and premolars expressed in the above formula has been made; but, judging from the condition in marsupials generally, I am induced to believe it to be correct. It is so given by Owen and Waterhouse.

Some fragmentary remains, consisting principally of jaws and isolated teeth, of a number of small mammals have been discovered from time to time in the Jurassic and Triassic deposits of this country, Europe, and South Africa, in which the teeth behind the canines reach as high a number as twelve in each lower jaw in some species. These are somewhat arbitrarily divided into an equal number of molars and premolars, but whether any or all of them had deciduous predecessors is not known. The reason for this division is that the first six behind the canine are premolariform in shape, while the others possess a number of sharp cusps. They have been referred to the marsupials and assigned a position near to *Myrmecobius*, but until their osteology is better known this is doubtful. Inasmuch as they are the oldest known mammals, we should anticipate on *a priori* grounds that they really belong to the monotremes instead of the marsupials. The great number of teeth certainly constitutes an approach to the Reptilia, and if they possessed a complete development of a second set, which is not at all improbable, the transition between reptile and mammal would be in a measure complete as regards the teeth.

Another strange and remarkable genus, *Plagiaulax*, together with a number of allies, comes from these ancient horizons. In this animal the molar pattern is complex for so early a representative of the Mammalia, and is difficult to understand. In the lower jaw of *Plagiaulax* there are seven teeth, of which the first is large, curved, and pointed, and is probably an incisor. This is followed after a considerable space by four teeth, all of which, except the first, are implanted by two roots and increase gradually in size. Their crowns are terminated superiorly by a wedge-shaped crest directed antero-posteriorly, which is rendered subserrate by the presence of a number of oblique vertical grooves. Behind these are two smaller teeth with tubercular crowns, which have been supposed to represent true molars.

The remaining marsupials which are really known to be such are divisible into the *Polyprotodontia*, or those of predaceous habits, having many incisors, and the *Diprotodontia*, vegetable feeders, having only two incisors, in the lower jaw. As far as dental characters go, they all agree in the possession of four true molars; there are never more than three premolars, and the deciduous molars, which are succeeded at a comparatively late period by the last premolars, are reduced to one in each jaw. This, therefore, furnishes another example wherein the definition of a premolar is violated.

Three families are included in the polyprotodont division, one of which, the opossums, is confined to North and South America, and the other two to the continent of Australia. As the common Virginia opossum is a good representative of this division, it is here taken for illustration and description. The dental formula is I. $\frac{5}{4}$, C. $\frac{1}{1}$, Pm. $\frac{3}{3}$, M. $\frac{4}{4}$ = 50. The incisors (Fig. 268) have a truncate cylindroid pattern,

Fig. 268.

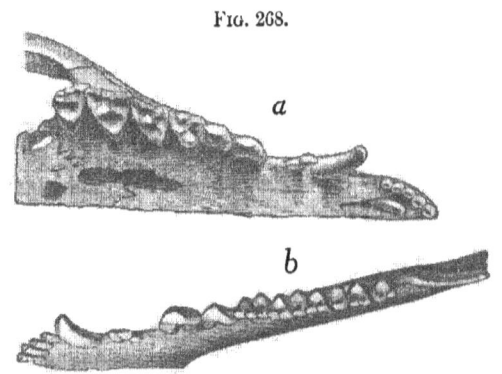

Dentition of Virginia Opossum (*Didelphis virginianus*): *a*, upper; *b*, lower jaw.

implanted by single fangs, and differ considerably from the corresponding teeth of the carnivores, which they exceed in number by two upon each side in the upper, and by one upon each side in the lower, jaw. The canines have relatively the same size and form as in the dog, and indicate clearly the carnivorous habits of their possessor. The premolars are simple premolariform teeth implanted by two roots, the first being smallest and separated from the other two by a diastema.

The molars of the lower jaw are essentially tuberculo-sectorial in pattern, with the external cusp of the anterior triangle largest. The heel is tritubercular and of large size. The molars of the upper jaw are interesting, inasmuch as they furnish a transitional stage in the formation of the W pattern described in the moles, shrews, etc. The first molar has the following structure: The crown is triangular in transverse section, with the apex directed inward, at which is situated the antero-internal cusp or the one corresponding with the single internal tubercle of the tritubercular molar. At the antero-external angle is situated a cusp of moderate dimensions, which in perfectly unworn specimens is more or less blended with the cingulum; just internal to this, upon close inspection, can usually be seen the rudiment of another cusp, which becomes better defined in the second molar. The exact homologies of these two cusps are not clear, but it seems very probable that the external is of cingular origin, and that the one internal to it is the true homologue of the antero-external cusp of the tritubercular tooth. On the outer edge of the crown, posterior to the two just described, is another cusp, which disappears in the last two molars, but which is well defined in the first and second. This cusp is homologous with the one which terminates the median external part of the W in the molars of the shrew and mole. A little posterior to a line drawn between this last-mentioned cusp and the one most internal is another large well-defined tubercle, from which a conspicuous ridge passes outward and backward to the produced postero-external angle of the crown.

It will thus be seen that all the requisite cusps are present in the first and second molars for the production of the W-structure, and that it would only require the presence of connecting ridges to complete it.

A distinctive characteristic of this, as well as most other marsupials, is seen in the strong inflection of the angle of the jaw and the vacuities caused by failure of ossification in the posterior part of the palatine bones.

Another family of this group includes the Phascogales, Tasmanian devil, the dog-headed opossum, etc. of the Australian continent and neighboring islands. This family is known to naturalists as the *Dasyuridæ*, and is distinguished from the opossums proper (*Didelphidæ*) by having the incisor formula $\frac{4}{3}$. In the genus *Phascogale* there are three premolars in the upper jaw and two in the lower; in *Dasyurus*, or the Tasmanian devil, there are only two premolars in each jaw, which number also obtains in the dog-headed opossum (*Thylacinus*).

The pattern of the molar teeth of this latter animal is very much like that of *Mesonyx*, consisting in the lower series of a principal cone, to which are added anterior and posterior basal cusps. The upper molars are tritubercular, as in that genus, but there is a considerable cingular ledge external to the two outer cusps.

The lower molars of the other two genera are very similar to those of the opossum, already described. The pattern of the upper molars of *Dasyurus* have been alluded to in connection with those of the shrew, and need no further description; those of *Phascogale* are essentially the same.

The bandicoots, constituting the family *Peramelidæ*, are distinguished by an incisor formula $\frac{5}{3}$. The canines are reduced, and placed relatively far back in the dentigerous border of the jaws. The molar and premolar formula is the same as in the opossum, and there is a similarity of pattern in the corresponding teeth of the two families.

In the second division, *Diprotodontia*, the incisors are reduced to two in the lower jaw; the canines are always small, and in many cases altogether wanting, while the molars are more complex, being better adapted to the mastication of a vegetable diet, upon which they principally feed.

The kangaroo furnishes a typical example of this group, and is here described. The dental formula of Bennett's wallaby (*Halmaturus bennetti*) is I. $\frac{3}{1}$, C. $\frac{0}{0}$, Pm. $\frac{1}{1}$, M. $\frac{4}{4} = 28$. The three pairs of incisors in the upper jaw (Fig. 269) are subequal and closely approximated, except in the middle line, where those of the opposite side are separated from each other by a considerable space. They have incisiform crowns, and are implanted by enlarged roots caused by an unusually thick coat of cement. These are opposed by a single tooth on each side below, whose direction is almost a continuation of the long axis of the jaw, so procumbent is its implantation. They are long teeth with enamel-covered crowns, slightly compressed from side to side, so as to present cutting edges on the surfaces which would correspond to the anterior and posterior faces if the tooth were erect, but which in its present position are superior and inferior. The superior edge bites against the three upper incisors, opposing them exactly.

After a long interval come the premolars, which have approximately the same structure in the two jaws as do the molars behind them. The premolars are implanted by two roots, and have crowns whose longitudinal diameter greatly exceeds the transverse. The summit of the crown terminates in an antero-posterior ridge, which is bordered at the base

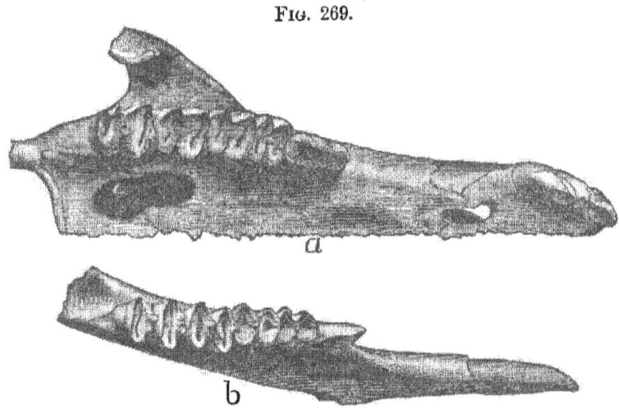

Fig. 269.

Dental Series of Kangaroo (*Halmaturus bennetti*): *a*, upper, *b*, lower jaw.

internally in the upper ones by a well-marked cingulum bearing several small cusps; this cingulum is absent from the inferior teeth.

The crowns of the molars are highly lophodont, consisting of two strong transverse crests connected in the median line by an antero-posterior ridge. They are all nearly equal in size and alike in both jaws.

In the phalangers, which constitute another family of this division, the incisors are the same as in the kangaroo. Small canines are usually present, and the premolars may be increased to three in the upper jaw. The third premolar has substantially the same structure as that of the kangaroos, but the molar pattern is selenodont, resembling in this respect the artiodactyle ungulates. They are quadritubercular, the four cusps being crescentic in section, with the crescents reversed in the lower jaw, just as in the artiodactyles.

Still another family is represented by the wombat, whose dentition exhibits a modification in the same direction as the rodent monodelphs in the reduction of the incisors to a single pair in each jaw and their growth from persistent pulps. The canines are absent, the premolars are $\frac{1}{1}$, and the molars, as well as incisors and premolars, grow continuously during the life of the animal. The molar pattern consists of transverse laminæ, greatly elongated and united by cement, much as in capybara, one of the rodents.

A gigantic extinct marsupial animal (*Thylocoleo*) has been described from the late Tertiary deposits of Australia, whose affinities and probable habits have provoked a good deal of discussion among English palæontologists. In each jaw there is a pair of enlarged, hooked, and pointed teeth in the position of the median incisors; these are followed in the upper jaw by three small teeth, the posterior of which probably represents a canine; in the lower jaw but a single tooth of this kind exists. Next follows a relatively enormous tooth, corresponding in pattern with the single premolar of the kangaroo, and is therefore trenchant. Behind these are one small tooth in the upper and two of like nature in the lower jaw.

From the trenchant nature of the large premolariform teeth, Prof. Owen, its describer, has considered it to have been carnivorous in habit, while Prof. Flower concludes, from the enlarged incisors and general resemblance of the enlarged teeth to that of the premolars of the kangaroos, that it is really affiliated with this group and was a vegetable feeder.

Other marsupials might be mentioned, but the principal modifications of the dental organs of this group have already been set forth in the types selected.

The Milk Dentition.

In the preceding pages we have spoken of the deciduous or milk dentition of the diphyodont Mammalia so far only as they relate to the permanent set in matters of definition. It now remains to discuss the more important question of their true nature and relationship to the permanent teeth in a philosophic sense. Are they superadded embryonic structures similar to the amnion and allantois, which subserve a temporary purpose and disappear with approaching maturity, or are they to be homologized with the first set of teeth of the lower Vertebrata?

Before proceeding to a discussion of these questions, it will first be necessary to give a general statement of the more important features of their anatomy, as well as the principal characters in which they differ from the permanent teeth.

As regards their development, it must be borne in mind that their enamel organs are originally derived from the lining membrane of the oral cavity, or at least that part of it which immediately covers the axes of the jaws, by a dipping down of the epithelium, while the dentine organ is developed from the underlying embryonic tissue. The enamel organs in this case are said therefore to arise *de novo*. After a time the enamel organs of the permanent incisors, canines, and premolars appear by a process of budding from the necks of the enamel organs of the deciduous teeth, but that of the first molar in the human subject arises *de novo*, just as those of the temporary teeth do from the primitive epithelial layer of the mouth.

From the neck of the enamel organ of this tooth the enamel organ of the second true molar buds out, while the third is derived from the second in a like manner. Whether this order of development is true of all diphyodont mammals is not known, and is a subject which very much needs further investigation.

The form of the milk teeth resembles that of the permanent ones which succeed them, as a general rule; an important exception to this, however, is to be observed in the last milk molar, which in the majority of cases is more complex than the permanent tooth which succeeds it. In the ungulates the last milk molar in the lower jaw resembles the last true molar in having three lobes, while in the upper jaw the last two milk molars have the complex pattern of the permanent molars. It is a rule of pretty general application that the last milk molar, and in many instances the last two, are succeeded by teeth of a simpler pattern. They may be well developed and retained in the jaw for a considerable period, as in the dog, or they may be extremely small, and shed, or rather absorbed, before birth, as in some of the seals. There may be as many as six in each jaw, as in the case of the nine-banded armadillo, or they may be reduced to a single one in each jaw, as in the marsupials. The usual number of milk molars is four in what may be called the typical diphyodont dentition, in which there are forty-four permanent teeth in all. Subtractions from this number are of common occurrence by reason of the first milk molar failing to develop a permanent successor or its complete disappearance. This, as we have seen, occurs in the dog and many other animals in which the number of premolars is normal. That this tooth is a persistent milk tooth is suggested by the fact that its enamel organ arises *de novo*, like those of the milk teeth generally.

In the monophyodonts one set has been lost, and the question naturally suggests itself, Which one is it? The very rudimentary condition of the milk teeth in the seals, which reaches an extreme point in the elephant seal, has led Prof. Flower to conclude that the single set of the monophyodonts is homologous with the permanent set of the diphyodonts, the first set having become rudimental and finally disappeared. He further concludes that the milk dentition generally is something superadded, and cannot therefore be homologized with the first set of the lower vertebrates. These conclusions are adopted by many authors.

In the first place, as regards the homology of the single set of teeth of the monophyodonts, there is much plausibility in Prof. Flower's position; but, upon the whole, our information respecting the exact limits of monophyodontism is too meagre to reach any satisfactory results in a solution of this question. It may yet turn out that many of the Cetacea, in which it is thought to be universal, really have rudimentary deciduous teeth in the early stages of growth, as has been suggested by Tomes. Among the edentates the nine-banded armadillo has already been cited as having two sets of teeth, and it does not seem at all improbable that all armadillos will ultimately be found to be diphyodont.

It should also be remembered that an approach to monophyodontism is made in many diphyodonts; and in all cases in which there is a partial loss of one set there can be little doubt that it is the *second* which has been subtracted. An example of this is afforded by the proboscidean series. In *Deinotherium* there were two and probably three permanent premolars; in some species of mastodons they are reduced in number to two or three; while in the existing elephants they have completely disappeared. The teeth which remain in the position of the premolars in these animals are certainly *persistent milk molars*. The first premolar of the dog, hippopotamus, and others is a case of the same kind. If monophyodontism has been produced in this way, then the single set which remains is not homologous with the permanent set of the diphyodonts, but combines the two, the molar dentition being made up of the true molars and persistent milk molars, with the permanent premolars subtracted.

With reference to the second conclusion, that the milk dentition is something superadded, Dr. Tomes very justly raises objection on the ground that the history of the development of the permanent teeth interposes a difficulty. He says:[1] "The tooth-germ of the milk tooth is first formed, and the tooth-germ of the permanent is derived from a portion (the neck of the enamel germ) of the formative organ of the milk tooth. Again, in most of those animals in which there is an endless succession of teeth, such as the snake, the newt, or the shark, each successive tooth-germ is derived from a similar part of its predecessor; the natural inference from which would be that the permanent set, being derived from the other, was the thing added in the diphyodonts."

Aside from the inherent improbability of this hypothesis of superaddition of the milk teeth, if the mammal has been derived from the reptile or batrachian—which is true if evolution is true—it is not at all remarkable, but, on the contrary, quite in keeping with the nature of the case, that the descendants should have retained some of their ancestral features. In the Batrachia and Reptilia there are many sets of teeth developed during the life of the individual, of which the first arises *de novo*, and all the succeeding ones are derived from that which precedes it. Altogether, I am disposed to regard the diphyodont mammalian dentition in the same light: those teeth which take their origin primarily from the epithelial lining of the mouth are strictly homologous with the first set of the lower vertebrates. This would include in the first set the deciduous incisors, canines, molars, and the first true or permanent molars. The second set of the batrachian and reptile would be represented by the permanent incisors, canines, premolars, and second true molar. The third succession would be represented by the last molar of the diphyodont dentition.

This view, of course, is based upon the presumption that the development of the true molars is the same in all diphyodonts as it is in the human subject—viz. that the enamel germ of the first is derived from the epithelial lining of the mouth; that that of the second is derived from the neck of the first; and that of the third from the second.

If it shall be found, however, on further investigation, that in any diphyodont the enamel germs of all the molars arise *de novo*, then they must in all such cases be added to the first set. This objection may be urged against the view that there are three, or even two, successions represented in the molars of the diphyodont—viz. that they do not succeed each other vertically, as in the case of the reptile and batrachian; but this I do not consider of vital importance. There is one thing upon which I would strongly insist, and that is that the first true molar in the human dentition is a persistent milk tooth.

[1] *Manual of Dental Anatomy*, p. 302.

CONCLUSIONS, ACKNOWLEDGMENTS, ETC.

THROUGHOUT the foregoing pages I have endeavored not only to give the leading characteristics of the principal modifications of the dental organs of the Vertebrata, but have in many cases, so far as our knowledge of the extinct forms would permit, endeavored to trace the leading steps in the production of the complex from the simple form. In so doing I have been made aware of the difficulties which beset such an undertaking: the principal burden of these difficulties lies in the comparatively imperfect knowledge we possess of the palæontological history of certain groups. In others the ancestry is more clearly indicated, and in my judgment the evidence is sufficient to demonstrate with a reasonable degree of certainty the more important steps in their dental evolution.

The modification of an organ from a simple to a complex structure necessarily implies a cause or force adequate to the production of such result. What, then, is the nature of the force or forces involved, and what is their method of operation? To simply say that this or that is so, that this tooth is simple and that is complex, without giving any reason why it is so, conveys little information. If one tooth is simple and another complex, there are reasons for it, and it is not only within the province, but is clearly the duty, of the odontologist to discover and point out these reasons if they can be found to exist.

Two explanations for all such phenomena have been offered. One of these presumes that they were created so by supernatural forces, but as to the nature of these forces we are not informed; much less do we know about the manner in which it was done. The other assumes that the natural or physical forces, operating through distinct and well-known physiological laws, are alone responsible for the resulting modifications.

Between these two explanations the naturalist experiences little difficulty in deciding which is most in accordance with the observed facts at his command. While the one rests solely upon the vaguest assumption, unsupported by so much as a single fact, the other rests upon observed scientific truth, which any one can verify who will take the pains to investigate. When we ascribe these modifications to the physical forces, the conclusion seems inevitable that those of a mechanical nature have been most largely concerned in the modification of form.

The change in form or size of any organ is principally due to addition, subtraction, or transposition of the histological elements of which it is composed; these, as is well known, are directly dependent on the amount of physiological waste and repair which the organ sustains, or, in other words, the extent of use and disuse. In proportion as an organ or a part of an organ is used, in that proportion will there be increased destruction of its substance and a corresponding determination of the nutritive fluids to supply the loss. The reverse is true of disuse.

In the harder tissues of the animal body strain and pressure have likewise been potent factors in the determination of form. Recognizing the importance of these influences, Mr. J. A. Ryder has constructed a most ingenious and far-reaching hypothesis in regard to the teeth, which he terms "the mechanical genesis of tooth-forms."[1] In this he satis-

[1] *Proceedings Acad. Nat. Sciences*, Philada., 1878.

factorily accounts for the forms and patterns of the molar teeth of the ungulates by the manner in which they have used their jaws. He has shown that in the bunodonts the mouth is simply opened and closed during mastication—a movement which is associated with a short-crowned tubercular molar—while in the selenodonts the lower jaw makes an extensive lateral sweep, and is associated with long-crowned crescentic molars. The conclusion is therefore obvious that as the bunodonts were compelled, through force of circumstances, to live upon a diet which required more extensive comminution before it could be properly assimilated, they gradualy develop greater mobility of the lower jaw; as a consequence of this, the patterns of the molar teeth were modified through pressure in accordance with this movement. If this proposition be true of the teeth of the ungulates, it must likewise be true of all other animals.[1]

Dr. Tomes in his *Manual of Dental Anatomy* criticises Mr. Ryder's conclusions, as follows: "The simple mechanical explanation that the teeth are drawn out into these forms hardly conveys much information, seeing that the tooth, before it is subjected to these influences, is quite finished, and its form, such as it is, is unalterable; while to effect an alteration in the form of a masticating surface an influence must be brought to bear upon the tooth-germs at an exceedingly early period. It might with equal justice be said that the crown of the tooth, being formed thus, had influenced the excursions of the jaw, and so modified the condyle."

It is evident that Dr. Tomes has either failed to grasp the meaning of Ryder's reasoning, or else denies one of the most important principles of the evolution doctrine. I am not aware that Ryder has anywhere asserted that the production of the selenodont pattern of the ungulate molar took place in a single generation, as Dr. Tomes's criticism would seem to imply. As a matter of course, the tooth of a modern ungulate when it comes into position is "quite finished," but were the teeth of the ancestors of the modern ungulates quite finished when they came into position? Ryder has attempted to show that this finishing process was a gradual one, which took many generations to accomplish, and the facts of palæontology bear out this view. The bold assertion of Dr. Tomes, to the effect that the masticating surface of a tooth when it comes into position is unalterable, is open to very grave and serious doubts. If the form of a bone or any other organ of the animal body can be influenced by impact and strain, as all evolutionists believe, then I can see no reason why a tooth is not amenable to the same influences.

The suggestion which Dr. Tomes offers, to the effect that the crowns of the teeth have determined the direction of the jaw movements, and so modified the condyle, is somewhat absurd. It is equal to assuming that structure has determined habit—a most remarkable conclusion for an evolutionist of the pronounced type of Dr. Tomes. The fact of the matter is, the evolution hypothesis assumes the very opposite of this. I have always believed it to be one of the cardinal

[1] Dr. C. N. Pierce has elaborated the views of Ryder and made important additions to this mechanical hypothesis.

principles of that great doctrine to consider that structure is largely the result of habit. Upon the whole, I find it quite impossible to harmonize such a suggestion with what this author holds on page 268 of the same work, in which he says : "It would be impossible in these pages to go through the arguments by which Mr. Darwin has established his main propositions; it must suffice to say here that he has fully convinced all those who are not in the habit, from the fixity of early impressions, of putting many matters upon another footing than that established by the exercise of reason, that any modification in the structure of a plant or animal which is of benefit to its possessor is capable—nay, is sure—of being transmitted and intensified in successive generations until great and material differences have more or less masked the resemblances to the parent form."

As a result of palæontological investigation we know that the form of the mandibular condyles has been very little, if any, modified, while the teeth have. We know, moreover, that it was a gradual process, and that all complex patterns had their origin in simple ones.

I feel well satisfied that there is not a single dentition of a complex nature that has not been profoundly modified by these same mechanical influences. If evolution has taken place as a result of the physical forces, it is impossible to discover any forces sufficient to produce such results other than those of strain, impact, and pressure. These have in some instances probably been exerted upon the young and growing tooth-germs ; in others they have operated upon the adult tooth, thereby furnishing the causes for individual variation and determining the direction of the hereditary energies.

In the preparation of the present article my grateful acknowledgments are due to the following gentlemen : to Prof. E. D. Cope of Philadelphia, who has kindly accorded me free access to his large and valuable collection of fossil vertebrates, without which it would have been impossible to include the extinct forms. He has likewise placed at my disposal all the illustrations in his possession which relate to his labors in this field. To Mr. J. A. Ryder for many wise and valuable suggestions in the developmental history of the teeth and other kindred subjects. To Dr. Theo. Gill for the loan of illustrations and much important information ; and, finally, to Prof. C. N. Pierce, at whose instance I was led to undertake the present work. I also wish to express my obligations to this gentleman for much kindly advice and assistance.

Of the works consulted I have made free use of C. S. Tomes's *Manual of Dental Anatomy*, a most useful and important work ; also, of the published writings of Profs. Owen, Huxley, Gegenbaur, Flower, Cope, Leidy, Allen, Ryder, Marsh, and others.

Descriptions of Plates.[1]

PLATE I.—Figs. 1 and 2 represent the deciduous incisors and cuspids, with their labial surfaces, and the molars with their buccal surfaces facing. Also the normal number of roots for these teeth in situ.

Figs. 3 and 4 represent the superior incisors, cuspids, and molars with their palatine surfaces, and the full inferior set with their lingual surfaces facing.

Figs. 5 and 6 represent the mesial surfaces of the full deciduous set, and both mesial and distal surfaces of the molars.

PLATE II.—Figs. 1 and 3 represent the full permanent set of thirty-two teeth, sixteen in each jaw, with the labial surfaces of the incisors and buccal surfaces of bicuspids and molars exposed: $a\ a$, the central incisors, right and left; $b\ b$, the laterals; $c\ c$, the cuspids or canines; $d\ d$, the first bicuspids; $e\ e$, the second bicuspids; $f\ f$, the first molars; $g\ g$, the second molars; $h\ h$, the third molars.

Figs. 2 and 4 represent the anterior teeth, incisors, and cuspids, with their cutting edges notched, as they are usually seen in the newly-erupted teeth, this uneven or notched appearance usually disappearing in a few months, or at most in a year, after eruption.

PLATE III.—Figs. 1 and 2 represent the deciduous or temporary teeth divided longitudinally through their lateral diameter.

Figs. 3 and 4 represent the same teeth divided through their antero-posterior diameters. These cuts give a very accurate idea of the relative size of the crown and roots, and of the position occupied by the pulp-chamber in the same.

PLATE IV.—Fig. 3 gives in contrast a sectional view of deciduous and permanent upper teeth divided through their lateral diameters.

Fig. 4, a sectional view of the corresponding lower teeth divided through their antero-posterior diameters. a, b, c, represent, respectively, the deciduous and permanent front incisors in contrast; d, e, f, the lateral incisors; g, h, i, the cuspids; k, deciduous molars, upper and lower; and l, m, the successors to the deciduous molars, the bicuspids; n, o represent permanent molars. c, f, i, m, o have dotted lines, indicating the thickness of enamel removed by wear, atrophy of the cementum, and reduction in the size of the pulp due to progressive calcification, these changes being incident to old age.

PLATE V. represents in Fig. 1, letters a to h and $\underline{a}$ to $\underline{h}$, the longitudinal or vertical sections of the sixteen superior teeth, showing the labio-palatine diameter of the pulp-chamber and canal in crown and roots, the section of the molars being through the anterior buccal and palatine roots, while the bicuspids $d\ e$ and $\underline{d\ e}$ illustrate the result of such a compression of the fang or root as to divide the pulp-chamber into two canals—a condition which so frequently exists in these flattened roots. The double-lettered series, $d\ d$ to $h\ h$ and $\underline{d\ d}$ to $\underline{h\ h}$, represent in the molars a section through the posterior buccal and the palatine roots, from which is quite readily recognized the slightly greater lateral diameter of the pulp-chamber in the crown and the larger canal in the posterior buccal root over that in the anterior buccal root, while the bicuspids lettered $e\ e\ d\ d$ and $\underline{d\ d\ e\ e}$ illustrate a modified pulp-chamber and canal, with bifurcation of the root in one, these being cut through a different axis or plane from the single-lettered series.

Fig. 2, letters a to h and $\underline{a}$ to $\underline{h}$, represents the sixteen inferior teeth with the section through their long diameters, as in the superior series. These incisors illustrate the compressed or flattened condition of their roots in contrast with the cylindrical character of the roots of the superior incisors, while the bicuspids $d\ e$ and $\underline{d\ e}$ illustrate the singleness of their pulp-chamber and the cylindrical condition of their roots as in contrast with the flattened or compressed condition of the roots of the superior bicuspids. The molars f, g, h, and $\underline{f, g, h}$ represent sections through the anterior root, illustrating its compressed condition and divided pulp-chamber in the first and second molar, and a somewhat flattened one in the anterior root of the third molar; $f f, g g, h h$, and $\underline{f f, g g, h h}$ represent the single and cylindrical pulp-chamber in the posterior root of the inferior molars, while $b\ b, c\ c$ and $\underline{a\ a, b\ b}$ represent the incisors and cuspids of the same series, with modified pulp-chambers arising from modified development.

PLATE VI.—Fig. 1, from a to h and $\underline{a}$ to $\underline{h}$, represents the superior teeth, with transverse or horizontal section through the base of the pulp-chamber in the crown, viewing the entrance to the canals of the several roots, while the same letters in Fig. 2 represent the inferior series in the same manner.

Fig. 3 represents the superior teeth, with the transverse or horizontal section made below the largest diameter of the pulp-chamber and through the canals after they have diverged from the central chamber, but before the roots into which they run have in the molars bifurcated.

Fig. 4 in like manner represents the inferior series, well illustrating the flattened or compressed condition of the canal in anterior roots of the molars and the division of the chamber, as is frequently found in the roots of the inferior incisors.

The letters $a\ a, b\ b, c\ c, d\ d, f f, \underline{d\ d}$ and $\underline{e\ e}$ (Fig. 3) represent the relative shapes, whether circular, oval, or flattened, of the pulp-canal in the roots of the superior central and lateral incisors, the cuspids, the first and second bicuspids, and the first, second, and third molars, while the same letters in Fig. 4 represent the relative shapes of the pulp-canal in similar teeth in the inferior series.

[1] These plates are taken from v. Carabelli's *Anatomie des Mundes*.

PLATE I.

For description, see page 504.

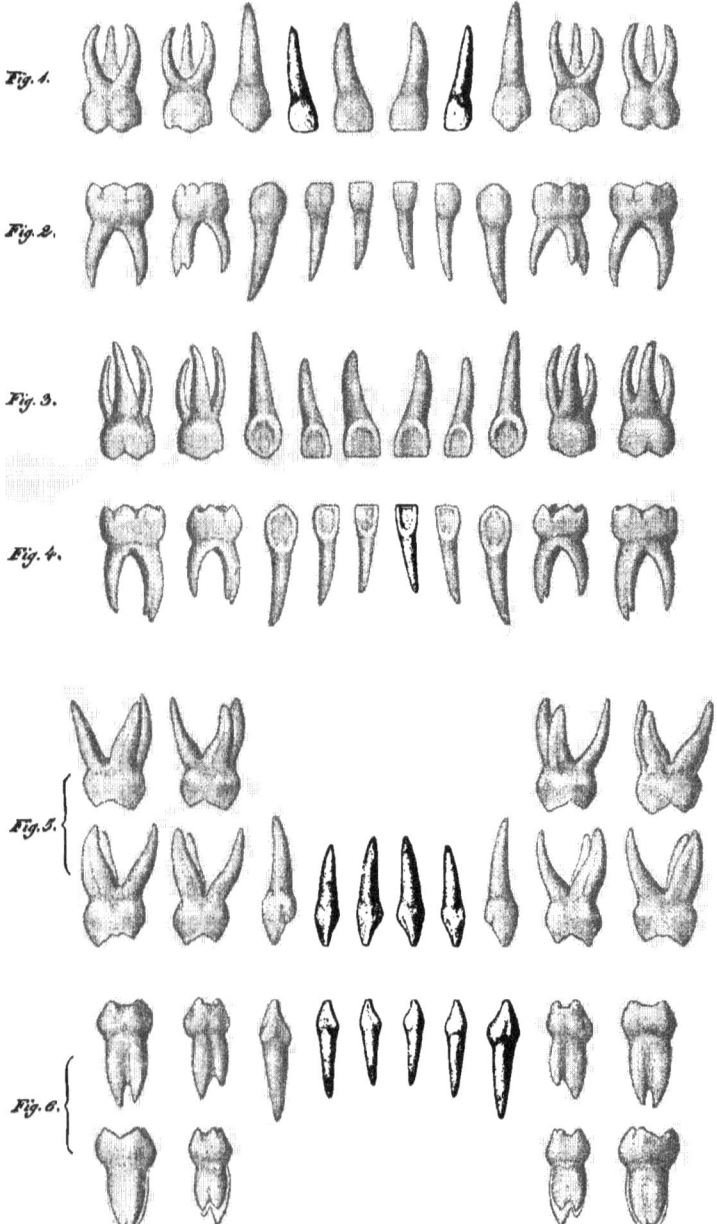

PLATE II.

For description, see page 504.

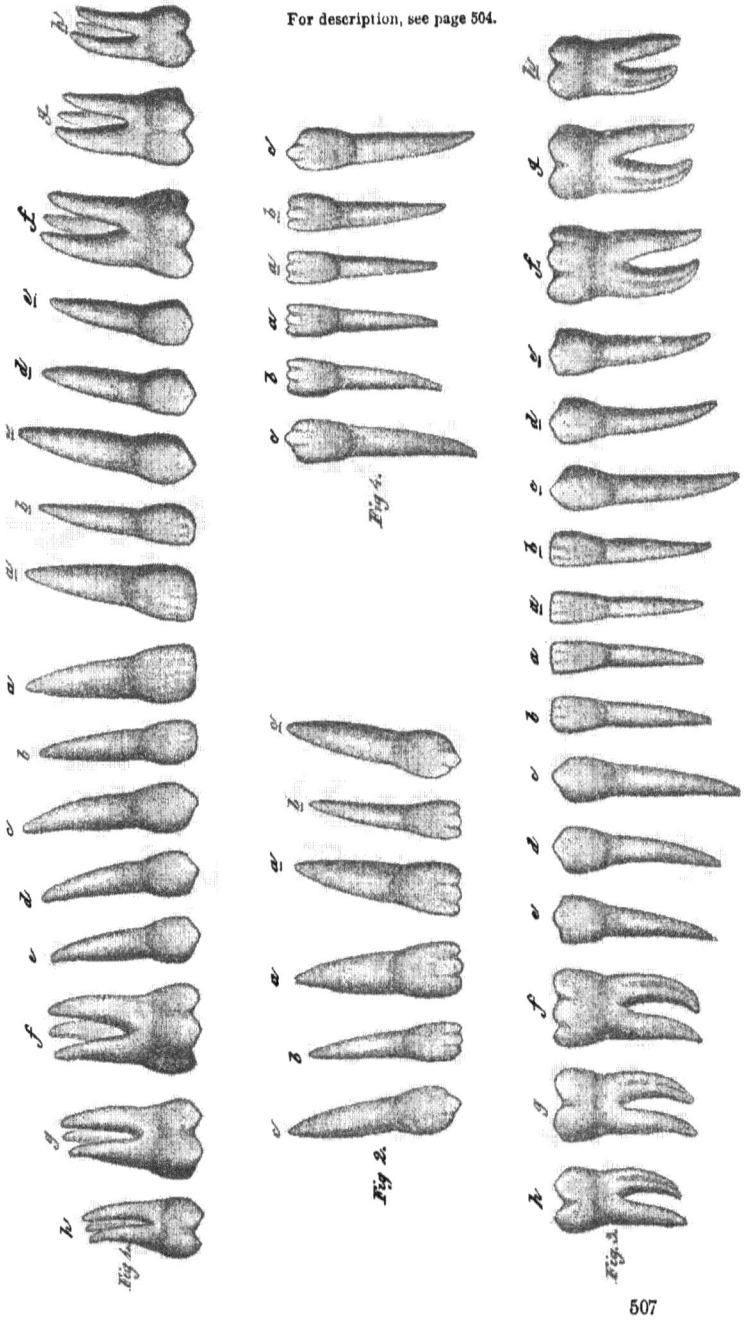

PLATE III.
For description, see page 504.

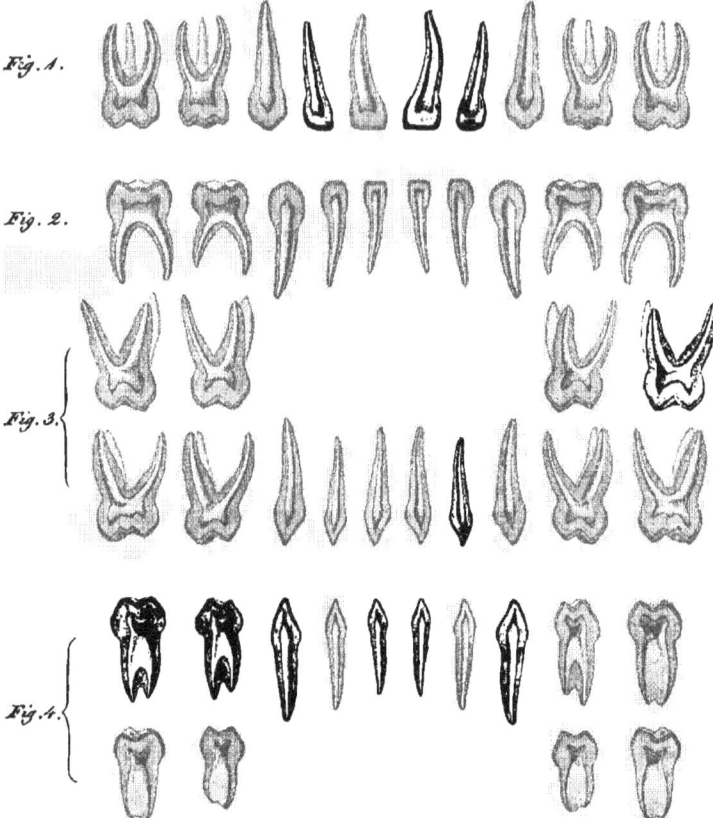

PLATE IV.
For description, see page 504.

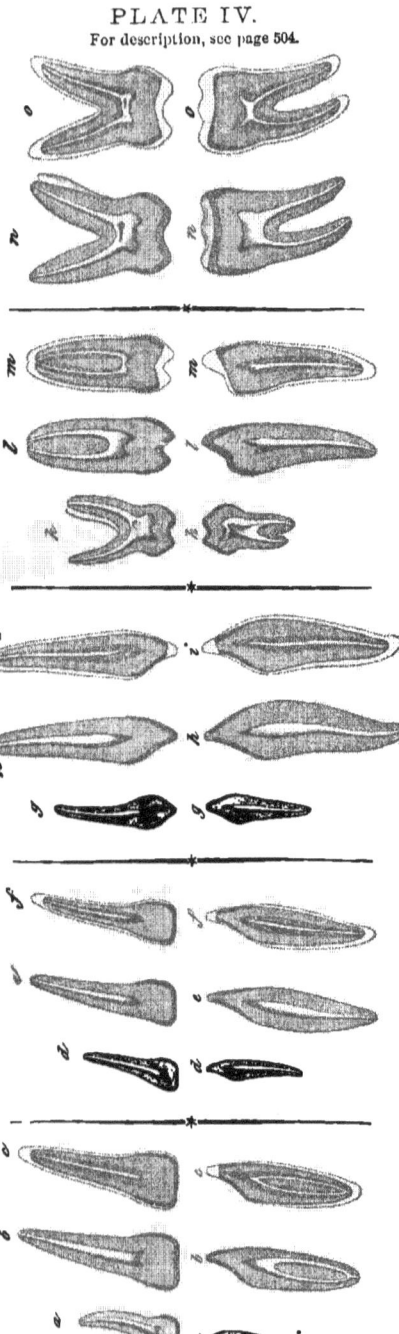

Fig. 3.   Fig. 4.

PLATE V.

For description, see page 504.

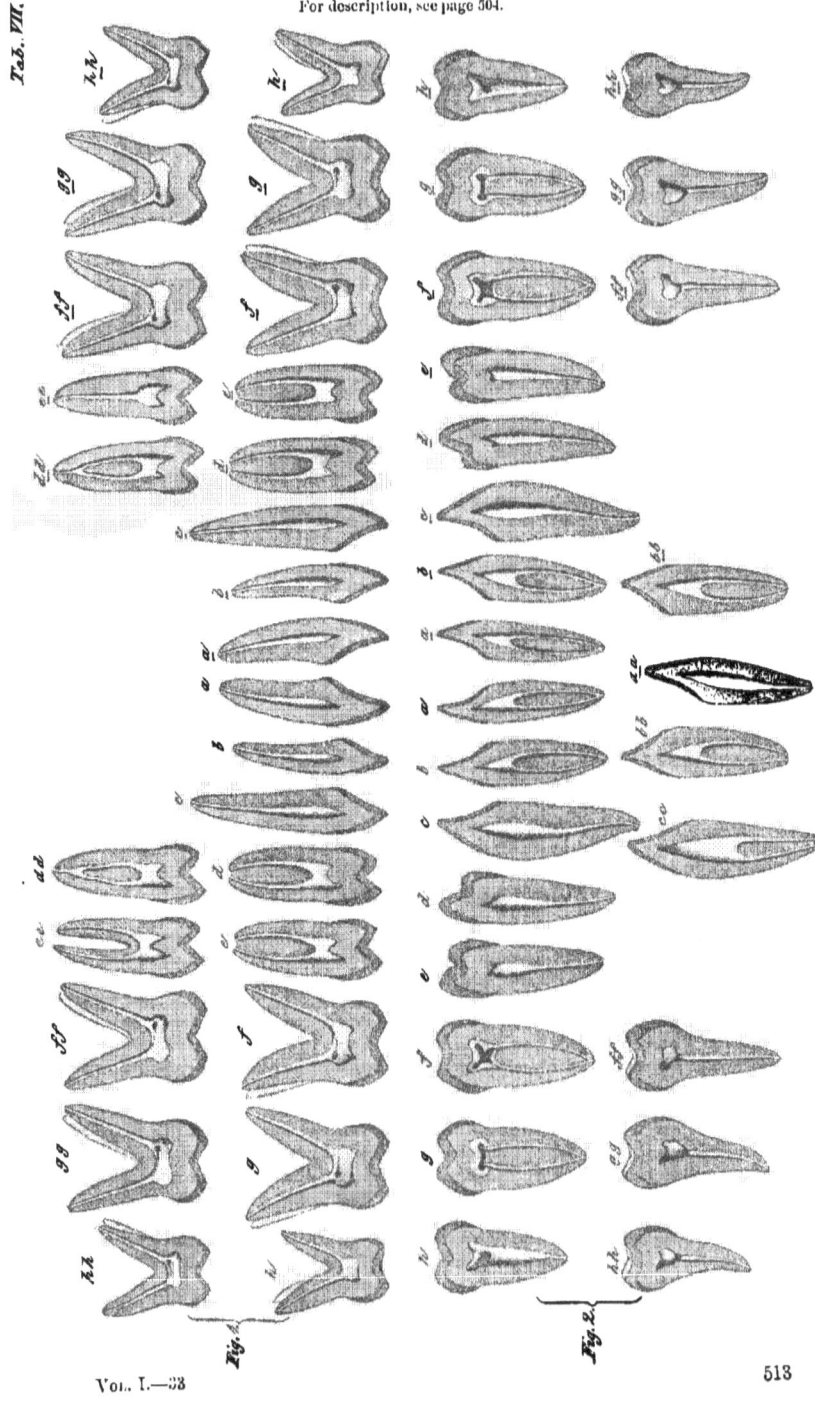

PLATE VI.
For description, see page 504.

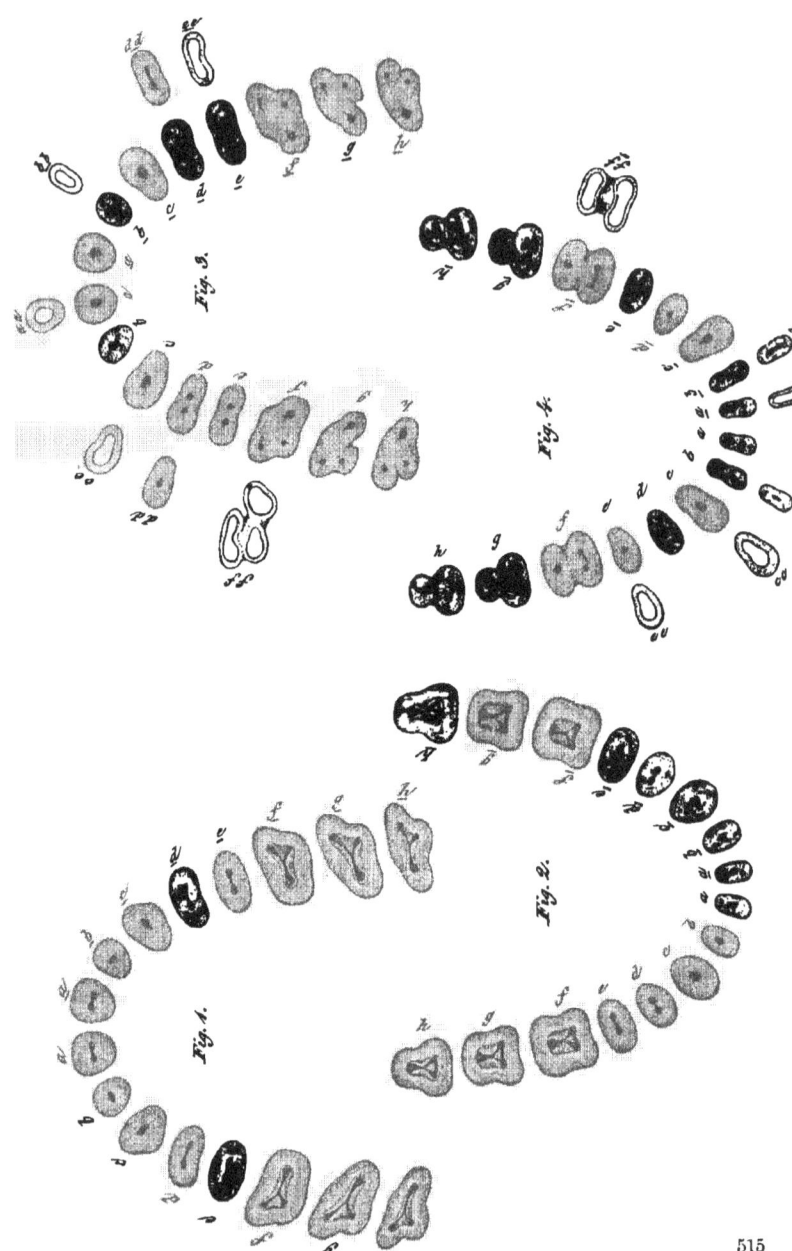

515

www.ingramcontent.com/pod-product-compliance
Lightning Source LLC
Chambersburg PA
CBHW022119160426
43197CB00009B/1082